Do Bats Drink Blood?

Animal Q&A: Fascinating Answers to Questions about Animals

Animal Q&A books invite readers to explore the secret lives of animals. Covering everything from their basic biology to their complex behaviors at every stage of life to issues in conservation, these richly illustrated books provide detailed information in an accessible style that brings to life the science and natural history of a variety of species.

Do Butterflies Bite? Fascinating Answers to Questions about Butterflies and Moths, by Hazel Davies and Carol A. Butler

Do Bats Drink Blood? Fascinating Answers to Questions about Bats, by Barbara A. Schmidt-French and Carol A. Butler

Do Bats Drink Blood?

Fascinating Answers to Questions about Bats

Barbara A. Schmidt-French
and Carol A. Butler

Rutgers University Press
NEW BRUNSWICK, NEW JERSEY, AND LONDON

Library of Congress Cataloging-in-Publication Data

French, Barbara A. Schmidt
Do bats drink blood? : fascinating answers to questions about bats / Barbara A. Schmidt-French, Carol A. Butler.
p. cm.—(Animal Q&A)
Includes bibliographical references and index.
ISBN 978–0–8135–4587–5 (hardcover : alk. paper)
ISBN 978–0–8135–4588–2 (pbk. : alk. paper)
1. Bats—Miscellanea. I. Butler, Carol A. II. Title.
QL737.C5F82 2009
599.4—dc22

2008048065

A British Cataloging-in-Publication record for this book is available from the British Library.

Visit our Web site: http://rutgerspress.rutgers.edu

Manufactured in the United States of America

Contents

Preface

Have you ever actually seen a live bat up close? For most people, the answer is probably no, yet many cringe at the very idea of seeing a bat. This reaction is probably based in part on horror stories and creepy movies that depict bats flying around at night causing mayhem. It is revealing that in countries where day-flying bats are common, they are considered lucky omens and are even thought of with affection. In general, bats in the wild are unthreatening, and their instinct, when disturbed, is simply to fly away. Like many other mammals and their offspring, some bats appear to be curious, affectionate, and even playful with one another.

Of the more than eleven hundred species of bats in the world, most of the forty-six species found in the United States are relatively small, comparable in size to a canary. A common American species, the insect-eating little brown bat (*Myotis lucifugus*), weighs only seven to nine grams, the combined weight of a nickel and a dime. The large bats in the popular imagination are Old World fruit bats, found in tropical areas of Asia, Africa, and Oceania.

Bats are highly beneficial animals. Large colonies of bats literally eat tons of insects every night, reducing the amount of pesticides that are needed on the crops farmers grow and, in turn, that end up in the foods we eat. Many bats in tropical areas pollinate plants and disperse seeds, making important contributions to crops, habitat maintenance, and rain-forest regeneration.

In the last few years, both traditional research and innovative projects using advanced technology have resulted in a flood of new and fascinating information about bats, their lifestyles, and their habitats. In this book we hope to satisfy your curiosity about bats, providing simple answers and including some of the latest scientific discoveries. It has been our goal to make the information accessible and enjoyable and to replace any misconceptions you may have with appreciation and curiosity. If you were already a bat groupie, we hope you will come to appreciate these fascinating creatures even more than you did when you opened this book for the first time.

Acknowledgments

We are grateful to the following people and organizations for allowing us to use their amazing illustrations and photographs for this book: the American Museum of Natural History in New York; Jesse Barber, Colorado State University; Alexander Baugh, University of Texas at Austin; Kirsten Bohn, University of Texas at Austin; Carol Bunyard and Gerald Carter, Cornell University; David Chapman; Michael Durham; Shawn Gould; Harvey J. D. Garcia, Conservation International, Philippines; Amanda Lollar, Bat World Sanctuary; Beverly Rivera-Walters; and Merlin Tuttle, Bat Conservation International. Thanks to George West for the bat drawings on the title page and to Donna Buonaiuto for helping us connect with the fossil.

We are grateful to Laura Redish, Native Languages of the Americas, for language translations and to Jesse Barber, Gerald Carter, Paul Cryan, and John Whitaker Jr. for their helpful suggestions as we wrote this book. The encouragement and support of our agent, Deirdre Mullane, and of our bat-enthusiast editor, Doreen Valentine, helped to make writing this book a rewarding experience. The authors enjoyed the opportunity to work together and to share our appreciation of bats with our readers.

Do Bats Drink Blood?

ONE

Bat Basics

Question 1: What is a bat?

Answer: Bats are the only mammals that can truly fly. They have elongated fingers that are connected by membranes to their torso, forming their wings. In common with all mammals, they have hair or fur covering their bodies and they are endotherms, which means they generate their own body heat instead of being dependent on the environment to regulate their body temperature. Baby bats, called pups, develop in utero and are born alive. They nurse on their mother's breast milk for the first few weeks of life. Most bats are relatively small. The largest bat is the size of a cat, while the smallest bat is not much larger than a bumblebee. They are second only to rodents with regard to the large number of species and the diverse range of their habitats.

A special characteristic of bats is that they hang in a head-down position. A bat's legs are rotated 180 degrees at the hip so that the feet point backwards, and they cling to their roost with the claws on their toes. Locking tendons in their toes allow them to hang from their feet without expending energy. Bats are clean animals that groom themselves regularly, and when they need to urinate or defecate in the roost, they turn themselves head-up.

Question 2: How are bats classified?

Answer: Within the animal kingdom (Animalia), bats belong to the phylum Chordata, which includes all animals with vertebrae,

Figure 1. *Lavia frons,* a monogamous yellow-winged bat, has blue-gray fur and yellow-wings. *(Photograph courtesy of M. D. Tuttle, Bat Conservation International, www.batcon.org.)*

and to the class Mammalia (mammals). Bats are the sole members of the order Chiroptera, from the Greek *cheir,* which means "hand," and *pteron,* which means "wing." The order Chiroptera is divided into two suborders: Microchiroptera, known as microbats, and Megachiroptera, or megabats. Microchiroptera are generally quite small, ranging in length from less than two inches to just over six inches (approximately four to sixteen centimeters), with most species at the smaller end of the range. They range in weight from less than one-tenth of an ounce to seven ounces (two to two hundred grams). To get a sense of the size of a representative bat of this type, consider the little brown bat (*Myotis lucifugus*), a common species in many parts of the United States, which weighs as much as the combined weight of a nickel and a dime.

There are seventeen families of Microchiroptera. Although there are a few exceptions to the rule, it is generally accurate to say that Microchiroptera use echolocation, or sound, to navigate and find food, and megabats do not. Most microbats feed on insects, although species in one family, Phyllostomidae, eat fruit or nectar, and a few others feed on small vertebrates. Some microbats have a good sense of smell, and most have relatively large ears that aid in echolocation. Some have unusual facial features that are thought to amplify sound, and although many microbats have relatively small eyes, those that feed on fruit or small vertebrates tend to have larger eyes.

Figure 2. *Macrotis californicus*, the California leaf-nosed bat, is an echolocating bat with big ears. *(Photograph courtesy of M. D. Tuttle, Bat Conservation International, www.batcon.org.)*

Figure 3. *Rousettus aegyptiacus,* the Egyptian fruit bat, has large eyes like most other megabats. It also uses a simple form of echolocation to navigate in dark caves. *(Photograph courtesy of Bat World Sanctuary, www.batworld.org.)*

All species of Megachiroptera belong to the single family Pteropodidae and are commonly called "flying foxes." Megachiroptera weigh from less than half an ounce to just over three pounds (ten to over fifteen hundred grams), most weighing between one-half to two pounds. They have prominent eyes, excellent vision, relatively small ears, and a good sense of smell. Megabats feed on fruit and nectar and they do not use echolocation to find food. Members of only one genus of megabats, *Rousettus,* use a rudimentary form of echolocation to navigate in caves. See also this chapter, question 5: When did bats evolve?

Question 3: How many species of bats are there in the world?

Answer: There are over eleven hundred species of bats in the world today, although this number continues to change as

Bats of the United States

Family Mormoopidae
Mormoops megalophylla, Peter's ghost-faced bat

Family Phyllostomidae
Artibeus jamaicensis, Jamaican fruit bat
Choeronycteris mexicana, Mexican long-tongued bat
Leptonycteris nivalis, Mexican long-nosed bat
Leptonycteris yerbabuenae, lesser long-nosed bat
Macrotus californicus, California leaf-nosed bat

Family Vespertilionidae
Antrozous pallidus, pallid bat
Corynorhinus rafinesquii, Rafinesque's big-eared bat
Corynorhinus townsendii, Townsend's big-eared bat
Eptesicus fuscus, big brown bat
Euderma maculatum, spotted bat
Idionycteris phyllotis, Allen's big-eared bat
Lasionycteris noctivagans, silver-haired bat
Lasiurus borealis, eastern red bat
Lasiurus blossevillii, western red bat
Lasiurus cinereus, hoary bat
Lasiurus ega, southern yellow bat
Lasiurus intermedius, northern yellow bat
Lasiurus seminolus, Seminole bat
Lasiurus xanthinus, western yellow bat
Myotis auriculus, southwestern myotis
Myotis austroriparius, southeastern myotis
Myotis californicus, California myotis
Myotis ciliolabrum, western small-footed myotis
Myotis evotis, long-eared myotis
Myotis grisescens, gray bat (myotis)
Myotis keenii, Keen's myotis
Myotis leibii, eastern small-footed myotis

(continued)

Bats of the United States, *continued*

Family Vespertilionidae, *continued*
Myotis lucifugus, little brown bat (myotis)
Myotis occultus, Arizona myotis
Myotis septentionalis, northern myotis
Myotis sodalis, Indiana bat (myotis)
Myotis thysanodes, fringed myotis
Myotis velifer, cave myotis
Myotis volans, long-legged bat (myotis)
Myotis yumanensis, Yuma myotis
Nycticeius humeralis, evening bat
Parastrellus hesperus, canyon bat
Perimyotis subflavus, tri-colored bat

Family Molossidae
Eumops glaucinus, Wagner's bonneted bat
Eumops perotis, greater bonneted bat
Eumops underwoodi, Underwood's bonneted bat
Molossus molossus, Pallas's mastiff bat
Nyctinomops femorosaccus, pocketed free-tailed bat
Nyctinomops macrotis, big free-tailed bat
Tadarida brasiliensis, Brazilian free-tailed bat

scientists use more sophisticated techniques to distinguish one species from another. Only about 20 percent of these are megabats; the other 80 percent are microbats.

Question 4: Where in the world are bats found?

Answer: Species of microbats live on every continent except Antarctica; they are also absent from a few remote islands in French Polynesia and the North Atlantic. Megabats live only in the Old World tropics, which refers to tropical areas of Asia, Africa, and Oceania.

Question 5: When did bats evolve?

Answer: Bats represent 20 to 25 percent of all living mammals and are the mammals with the second largest number of species after rodents. However, they have left very few representations in the mammalian fossil record, and many of the fossils are incomplete; one that has survived, for example, consists of only a single tooth. In 2008, a report was published by Nancy Simmons of the American Museum of Natural History and her colleagues describing the analysis of two fossils that represent a new species of bat from the early Eocene era (approximately 52 million years ago). Found in Wyoming in 2003, an almost complete skeleton was beautifully preserved in the fine sediment of a fossil lake. The new bat (*Onychonycteris finneyi*) has features that are more primitive than have been seen in other early specimens, and scientists have concluded that this is the oldest species of bat that has ever been found. Analysis of its anatomy indicates that it was an insect-eater, capable of powered flight but not of echolocation. Its limbs and claws suggest that it probably was a good climber and was able to hang from a tree branch, much like present-day bats that roost in trees.

This new finding supports the hypothesis that the ability to fly in bats evolved before the ability to echolocate, a subject of controversy for many years because the earliest fossils known prior to this discovery represented bats that could both echolocate *and* fly, and there are no fossils of a transitional species between bats and their non-flying ancestors. Evidence that flight evolved before echolocation has major significance.

Karen Sears at the University of Colorado studied the development of the elongated fingers or digits that provide the skeletal support for the wings of a bat. She found the gene that controls their specialized growth, and when she applied the protein produced by this gene to the digits of a mouse embryo growing in her lab, its digits elongated just like the digits of a bat. If the same dramatic results observed in the mouse occurred when this gene became activated in the ancestors of today's bats, its activation might explain the absence of transitional species.

Figure 4. *Onychonycteris finneyi,* a fossil bat found in Wyoming, is the oldest species of bat that has ever been discovered. It was an insect-eater, capable of powered flight but not echolocation. *(Photograph courtesy of American Museum of Natural History, New York.)*

The order Chiroptera has been divided into two suborders and four superfamilies. Multiple genes have been sequenced from representatives of all the bat families, and molecular evidence now suggests that the ancestors of present-day bats originated from a single order (monophyly) that evolved at least by the late Paleocene era (65.0 to 54.8 million years ago), after the dinosaurs became extinct. It is believed that bats evolved on the supercontinent of Laurasia, which included most of today's northern continents, but the recent discovery of six new species of late Eocene (37 to 34 million years ago) fossils by Gregg Gunnell and his colleagues in northern Egypt has added to the mystery. Finding the ancestors of modern species of microbats in Africa suggests that primitive bat species may have dispersed there along with the primates about 50 million years ago and then later developed into the modern species that have dispersed across most of the world.

Question 6: Are all bats black?

Answer: Although many bats are shades of black, gray, or brown, there are others that are orange, red, or yellow in color. Female red bats (*Lasiurus borealis*) have beautiful orange-red fur, and a mother with her wings and fuzzy tail tucked around her babies resembles a piece of fruit hanging in a tree. Some of the flying foxes have bright red, orange, or yellow collars around their necks that are often more prominent in males. The very unusual-looking spotted bat (*Euderma maculatum*) has black fur with three white spots on its back. Most surprising are bats that are actually white. White bats (*Diclidurus*), also known as ghost bats, live in Latin America. Little Honduran white bats (*Ectophylla alba*) are white with yellow ears and nose leaf (see color plate B).

Although few bats are totally white, there are rare albinos with pink eyes (see color plate H). The fur on the underside of many bats is paler than the fur on the back, and some bats have fur that is dark at the base of each hair, but pale on the tips. Hoary bats (*Lasiurus cinereus*) have beautiful, luxuriant fur that is frosted at the tips with silver. The tri-colored bat (*Perimyotis subflavus*) has tri-colored fur that is dark at the base of each hair, yellow-grey in the middle, and dusky again at the tip.

The skin on the wings and ears of many bats is black or dark in color, but others have pale skin. For example, the ears and wings are yellow on the African yellow-winged bat (*Lavia frons*), the ears are pink on the spotted bat (*Euderma maculatum*), and the edges of the ears of the hoary bat (*Lasiurus cinereus*) are ringed in black. The pale yellow canyon bat (*Parastrellus hesperus*) looks like it is wearing a black mask, and the pale fur around the eyes of the spectacled bat (*Pteropus personatus*) makes it look like it is wearing glasses. Markings often help to camouflage bats, for example, proboscis bats (*Rhychonycteris naso*) have pale zigzags on their back that make it difficult to distinguish them from the tree bark where they roost. Several species, like the Jamaican fruit bat (*Artibeus jamaicensis*), have pale vertical stripes on their faces that help them blend into vegetation where they feed.

Males of some bat species are brighter than the females, and some have bright tufts of fur on the head or shoulders that attract females for mating. For example, the male Chapini's crested free-tailed bat (*Chaerephon chapini*) has a strip of long hairs on the top of its head that stands erect like a Mohawk haircut.

Figure 5. In 2007, Jake Esselstyn and a team of researchers discovered the Mindoro striped-faced fruit bat (*Styloctenium mindorensis*) on the island of Mindoro in the Philippines. *(Photograph courtesy of H. J. D. Garcia.)*

Question 7: How long do bats live?

Answer: The longest known life span of a bat is about three and a half times longer than the life span of other mammals of similar size. This is likely influenced by their low reproductive rate and their relative lack of vulnerability to predation. For example, rodents like mice or shrews only live three or four years at the most, while the record for the oldest bat is over forty-one years. This was a male Brandt's bat (*Myotis brandtii*) captured in Siberia. When the bat was first caught, a band with an identifying number was put on its forearm, which allowed researchers to identify it when it was recaptured forty-one years later. There are now records of six species that have lived for more than thirty years, and twenty-two species that have lived for more than twenty years in the wild.

Researchers have found that the life span of a bat is affected by its reproductive rate, so that bats that give birth to multiple pups each year may not live as long as bats that give birth to only one pup annually. Bats that hibernate tend to live a few years longer than bats that don't hibernate, perhaps because hibernation reduces the likelihood of starvation (see chapter 3, question 9: Do bats hibernate?). Also, cave roosts minimize the exposure of bats to predators, and the relatively stable temperature in a cave limits exposure to extreme heat or extreme cold. It has been established that caloric restriction increases longevity in some other mammals, and researchers speculate that this element of hibernation may also contribute to the longevity of bats.

Question 8: Which is the biggest bat?

Answer: Most of the largest bats in the world are fruit-eating Megachiroptera (megabats). The Philippine flying fox (*Pteropus vampyrus*) and the golden-crowned flying fox (*Acerodon jubatus*) can weigh two pounds (one kilogram) or more and have a wingspan of more than six feet (two meters). Another heavyweight is the Indian flying fox (*Pteropus giganteus*), which can weigh up to three and one-half pounds or about one and one-half kilograms. The Philippine fruit bat lives in Thailand, Indochina, Tenasserim, Malaysia, Indonesia, and the Philippines. The golden-crowed flying fox also lives in the Philippines, and the Indian flying fox is found in Pakistan, India, Nepal, Sikkim, Bhutan, Myanmar, Sri Lanka, and the Maldive Islands.

Question 9: Which is the smallest bat?

Answer: The smallest bat in the world is Kitti's hog-nosed bat (*Craseonycteris thonglongyai*). This insect-eating bat is so small that it is known as the "bumblebee bat," weighing about as much a penny. Kitti's hog-nosed bat is an endangered species found only in Thailand and Myanmar.

Question 10: How far can bats fly?

Answer: Most microbats feed within six to nine miles (ten to fifteen kilometers) of their daytime roost, although some fly as far as fifty miles (eighty kilometers). Large megabats may travel more than thirty-one miles (up to fifty kilometers) from their day roost to feed. The longest distances some bats travel occur when they migrate between winter and summer roosts (see chapter 3, question 8: Do bats migrate?). The distance traveled during migration varies from one species to another, with some traveling more than a thousand miles. There is even a record of a noctule bat (*Nyctalus noctula*) traveling from southern Russia to Greece, a distance of over twelve hundred miles (two thousand kilometers).

Question 11: How fast do bats fly?

Answer: Flight requires upward force (lift) and forward force (thrust). Lift occurs when air flows faster over the upper surface of the wing than it does over the lower surface; this allows the bat to overcome gravity and stay in the air. Thrust counters the effects of friction as the bat moves along. When a bat takes flight, the wings are first extended above its body and slightly

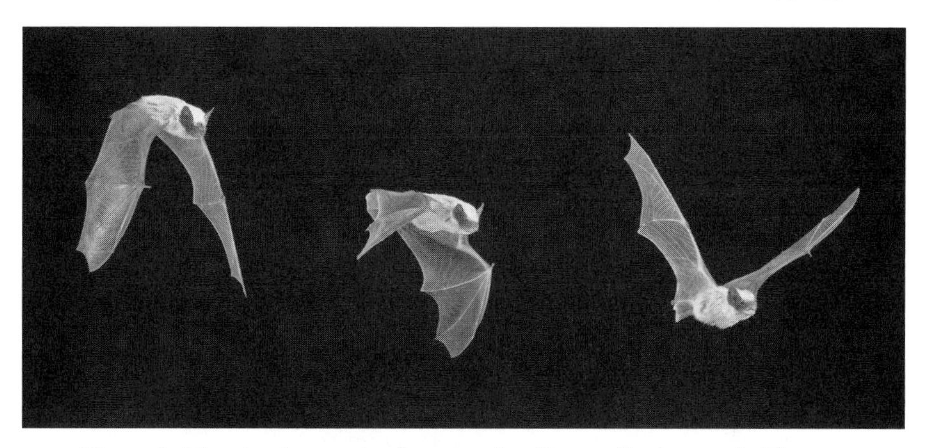

Figure 6. The wing-beat cycle of a canyon bat (*Parastrellus hesperus*). (*Photograph courtesy of Michael Durham, www.DurmPhoto.com.*)

backwards. Next the wings move down and forward in a downstroke. Then the wings fold slightly during an upstroke, where they are again extended above the bat's body and slightly backwards. Combined, this series of movements is called a wing-beat cycle. Bats fly by repeating this cycle over and over again. This link is to a video of a bat flying in a wind tunnel, where you can observe its movements in detail: http://media.newscientist.com/data/images/ns/av/dn11105V1.mpg.

Some bat species fly much faster than others, depending on the relationship between the area of the wing and the wing span (length from wing tip to wing tip), referred to as the *aspect ratio*. Bats with long, narrow wings are generally fast fliers that feed in open habitats, like the Brazilian free-tailed bat (*Tadarida brasiliensis*). These bats fly at an average speed of twenty-five miles per hour (forty kilometers per hour) but can approach speeds of forty-seven miles per hour (about seventy-five kilometers per hour) in level flight, and with a tail wind they can reach amazing speeds—over sixty miles per hour (over ninety-six kilometers per hour). Bats with short, broad wings are slower fliers, typically maneuvering through vegetation or other cluttered environments, plucking insects from leaves or hovering to drink nectar from flowers.

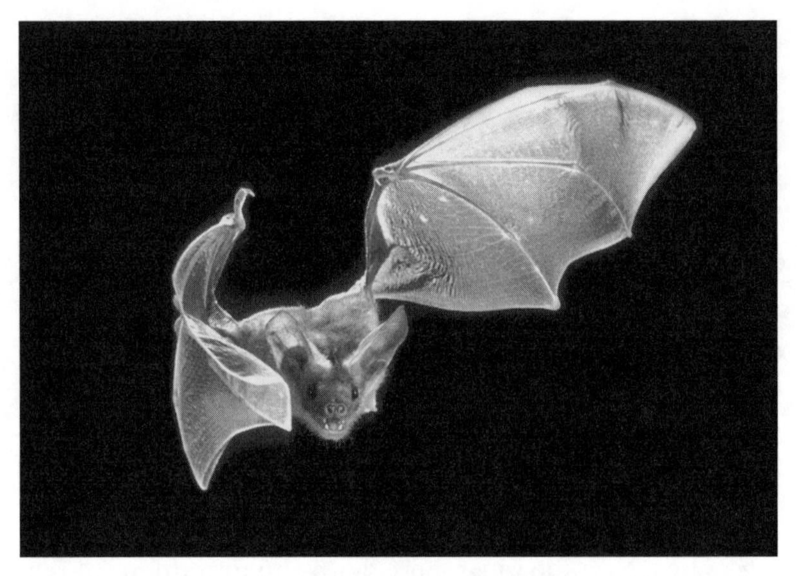

Figure 7. The pallid bat (*Antrozous pallidus*) is an agile flier with broad wings. *(Photograph courtesy of Michael Durham, www.DurmPhoto.com.)*

Question 12: How high do bats fly?

Answer: Some bats fly very low while foraging for food, taking insects or other prey directly from the ground, like the pallid bat (*Antrozous pallidus*), which ranges from southwestern Canada to central Mexico and Cuba. In contrast, the fast-flying Brazilian free-tailed bat (*Tadarida brasiliensis*) has been documented flying at altitudes of nearly ten thousand feet (over three thousand meters), higher than any other species.

Researchers found that the Brazilian free-tailed bats' feeding activity was correlated with the northward flight of huge waves of corn earworm moths (*Helicoverpa zea*), also known as cotton bollworm moths and tomato fruitworm moths. In the first few weeks of June, these moths emerge from the Lower Rio Grande Valley of Mexico and rise to hundreds of feet above the ground, using the winds to help them travel north to lay eggs on newly emerging crops. Three weeks after the eggs are laid, the larvae hatch and feed on the crops in south Texas. In short order, they pupate and then complete their growth cycle by *eclosing* (emerging from their pupae, having metamorphosed from larvae into adult moths) and flying further north to lay their eggs. This cycle repeats with new generations of moths continually moving northward, following the warm weather and the sprouting crops, and followed by the bats. Corn earworms are pests on a tremendous variety of crops including corn, cotton, tomato, artichoke, asparagus, cabbage, cantaloupe, soybean, sugarcane, and many others.

Researchers at the Mexican border floated weather balloons high among the moths, and radio microphones in the balloons recorded bat calls at altitudes as high as thirty-nine hundred feet (nearly twelve hundred meters), verifying that the bats were flying along with the moths. To confirm that the bats were indeed feeding on the moths, Gary McCracken's laboratory at the University of Tennessee in Knoxville developed a DNA marker that allowed them to identify gene fragments from these moths in the bats' feces. The results confirmed that the bats were indeed feeding on the moths.

Question 13: Are bats intelligent?

Answer: Learning and memory are components of intelligence, as is the ability to use tools. Complex social behavior and the ability to communicate are also associated with higher levels of intelligence.

An important theory regarding the evolution of intelligence relates feeding strategies to intellectual development. Extractive foraging, or locating and processing embedded foods, is considered evidence of a higher level of intelligence, and this behavior is typically attributed to primates that use tools to accomplish these tasks. According to Barbara King (College of William and Mary in Virginia), "some acts of extractive foraging by nonprimates are equally sophisticated as those of primates." Peeling the skin off a mango in order to prepare it to be eaten is described by Natarajan Singaravelan (University of Haifa at Oranim in Israel) and Ganapathy Marimuthu (Madurai Kamaraj University in India) as a form of extractive foraging, and they observed this behavior in short-nosed fruit bats (*Mangifera indica*).

Echolocating bats can adapt their calls to environmental cues, varying the frequency, duration, and bandwidth of pulses depending on feedback from obstacles and prey. This is evidence of a more evolved "technical" intelligence in many species of bats, according to Kamran Safi and colleagues at the University of Zurich in Switzerland. Bats that have large, broad wings relative to the size of their body and that live in dense environments with many obstacles tend to be highly maneuverable flyers that require a lot of energy to forage successfully. If they are echolocators, they need good spatial memory and their hearing is likely to be particularly acute. The brains of these bats tend to be larger than the brains of species that forage in open spaces. Bats that forage in open spaces tend to have smaller, narrower wings relative to body mass, and they rely on speed rather than maneuverability, so flying for them is more efficient and less costly. Their sensory needs are reduced, and having a smaller brain gives them the advantage of reduced weight, improved aerodynamics, and lowered energy costs.

A variety of experiments have shown that pollinators can learn to associate colors, visual images, and even sound with food sources. While these are neat experiments, they also have important implications for higher-level functioning in an environment that is always changing. Butterflies and bees as well as bats can learn which flowers are good nectar sources, and they all feed more quickly and efficiently on subsequent visits to the same flower, even if it has a complex shape. Bats have spatial memory, returning directly to a good nectar source without having to rediscover it each time they forage—again, evidence of a relatively high level of functioning in being able to retain information of this complexity.

European free-tailed bats (*Tadarida teniotis*) feed on moths and other common garden insects that have some capacity to hear the bats' echolocation calls and avoid being captured. However, Jens Rydell (University of Aberdeen in the United Kindgom) and Raphael Arlettaz (University of Lausanne in Switzerland) found evidence that when bats hunted for those insects, they used echolocation calls that the insects were unable to hear because the frequencies were above or below the insects' hearing range. The bats had adapted to the insects' defenses.

Rachel Page and Michael Ryan at the University of Texas in Austin conducted experiments with fringe-lipped bats (*Trachops cirrhosus*) and found that the bats could learn by observing their neighbors' behavior. What is more interesting, they rapidly learned to eat in association with sounds that they would normally avoid in the wild, indicating an impressive degree of adaptability and flexibility. These are bats that normally eat frogs, and in the experiments the calls of toxic toads were played for captured bats in large outdoor flight cages. When they approached the loudspeaker that was playing the sounds, they were rewarded with food. They quickly learned to associate the sound of the toxic frog with food, even though in the wild that sound would warn them away from the prey. When newly captured bats were allowed to observe these bats, they learned to associate the toxic toad call with food after observing the behavior an average of just five times. Rachel Page inadvertently recaptured an old and distinctive *Trachops* bat that she had captured a year earlier and had used in another experiment. After a one-year hiatus in the wild, the old bat remembered the routine and repeatedly came to take small fish from Page's hand when she made the clicking sound the bat had been conditioned to the previous year.

There are experiments that indicate that bats can communicate with one another and can make group decisions, sometimes contrary to what experience would predict. Female greater spear-nosed bats (*Phyllostomus hastatus*) live in colonies, and they call to one another to coordinate group foraging. Gerald Kerth at the University of Zurich in Switzerland conducted experiments in which he gave wild Bechstein's bats (*Myotis bechsteinii*) conflicting information about the suitability of roosts in their habitat in order to study how they make group decisions. The bats live in small colonies in tree cavities and bat boxes (constructed bat houses placed in the habitat), and they change roosts frequently, apparently deciding at night where to roost next. Forty-four bats were marked so that their behavior could be tracked, and a batch of new bat boxes was installed in the habitat. Individual bats often visited a new box several times before using it as a roost, and at first the entrance to the roost was left invitingly open so the bats could enter. After a few bats from the colony had visited the box and departed, the experimenter blocked the entrance to the box with wire mesh so the next visitors could not enter and would find the same location unsuitable as a roost. Another experiment was designed to give some bats an alarming experience at a roost that others had recently experienced as safe. Waiting until some of the bats had emerged uneventfully at dusk from a particular roost, the experimenter then sought to alarm the remaining bats by making scratching sounds on the outside of the box with aluminum foil attached to a stick. The observers found that each individual bat's experience of a roost as suitable or unsuitable did not completely predict where it would choose to roost. There were times when the experience of a significant number of bats in the group led to a group decision to use the roost, even if some of the bats in the group had experienced it as unsuitable. At other times, when a group of bats had conflicting information, they split into smaller groups rather than acting together, but when the members of a group were given the same information, the group acted cohesively.

Amanda Lollar and Barbara Schmidt-French documented an elaborate array of social calls within captive colonies of the gregarious Brazilian free-tailed bat (*Tadarida brasiliensis*). Kirsten Bohn, Ta-Sheng Ma, and associates at the George Pollak bat lab at the University of Texas in Austin, together with Barbara Schmidt-French, further documented the assimilation of some calls into more complex units that may demonstrate a simple use of syntax, another indication of higher-level intelligence. So, are bats smart? It does seem that they have an impressive capacity to learn, an exquisite sensitivity to their environment, and the ability to communicate with each other and to make group decisions, all indications that they have a relatively high level of intelligence.

Question 14: Do bats drink blood?

Answer: Only three species of small bats actually drink blood. These are the true vampire bats, each weighing less than two ounces. They include the common vampire (*Desmodus rotundus*), the hairy-legged vampire (*Diphylla ecaudata*), and the white-winged vampire (*Diaemus youngi*), all native to Mexico, Central and South America, and the Caribbean. The hairy-legged vampire and the white-winged vampire feed mainly on blood from birds. Gerald Carter from Cornell University used a noninvasive method to investigate the diet of vampire bats that fed on birds. He extracted bird DNA from the bats' droppings and was able to accurately determine the species of birds on which they preyed.

The common vampire feeds mainly on the blood of mammals such as cattle, pigs, or goats. If a colony of common vampire bats has been feeding on a large herd of cattle that is suddenly sold off and moved, some of the bats may feed on humans if alternative food sources are not available, but they choose individuals to whom they have easy access, like people who sleep outdoors or in homes with no screening on the windows.

Vampire bats use echolocation to orient themselves during flight, but they probably rely on many factors, including smell and spatial memory, when choosing individual prey, allowing them to recall an environment and to return to the same feeding location without extensive searching. Once a vampire bat has chosen its prey, it typically lands on the back of the animal or approaches it from the ground, usually when the animal is asleep. Using heat-sensitive pits in its nose, the bat finds a place on the prey where there is a good blood supply just under the skin and licks the spot for several minutes, softening it before making a small cut with its razor-sharp incisors. After latching onto its prey, an anticoagulant in the bat's saliva is channeled down a groove on the underside of the bat's tongue and into the wound. The bat then laps up the blood, which flows from the wound along a groove on the upper side of the bat's tongue and into its mouth (see color plate E).

Bat Breath

Co-author Barbara Schmidt-French has cared for hundreds of bats in her years as a bat rehabilitator and finds that a bat's smell is sometimes helpful in diagnosing an illness. For example, bad breath may indicate a dental infection. Scientists are now paying attention to bat breath to determine what vampire bats last ate.

Vampire bats in Costa Rica feed on rain-forest mammals like tapirs and peccaries, but also feed on the blood of cattle. The rain-forest mammals feed on different plants than those on which cattle feed, and the plants can be distinguished by their carbon *isotopes* (chemical compounds). By analyzing the isotopes in the carbon dioxide a bat exhales as it breathes, scientists are able to determine the animal on which the vampire bat recently fed.

When Christian Voigt and his colleagues analyzed the breath of vampire bats in Costa Rica, they found that the last blood meal of most of them appeared to have originated from cattle, although some bats had fed on blood from rain-forest mammals. Scientists do not believe the bats prefer the blood of cattle, but rather that cattle are often fenced in open pastures where they are easily accessible to the bats, while rain-forest mammals are more likely to hide in dense vegetation. As rain forests are cleared for cattle ranching in Latin America, common vampire bats (*Desmodus rotundus*) have become more dependent on cattle than native forest animals for blood meals. The conversion of rain forests to pasture land for cattle ranching has caused an increase in populations of the common vampire bat in Latin America.

Researchers have explored some of the elements that attract a vampire bat to its prey. In one experiment conducted by Udo Gröger and Lutz Wiegrebe (Ludwig Maximilians University in Munich, Germany), two vampire bats were each taught to associate a recording of a different person breathing with a particular cattle blood dispenser that rewarded them with a blood meal. The bats were then played short clips of the breathing, and they went unerringly to the dispenser associated with the particular breathing sound they had learned to identify with food, recognizing the breathing whether it had been recorded when the person was resting or after exercise. Joseph Bahlman and Douglas Kelt (University of California at Davis) conducted what they described as "a modified cafeteria trial" in which bats were offered blood meals associated with the scent of fur or feces from cows and other meals that had no olfactory cues (associated smells). Bats showed a significant preference for the blood associated with the cows' scents. These experiments suggest that, along with spatial memory and vision, other sensory clues play a role in the ability of vampire bats to return to the same prey night after night.

Blood is about 90 percent water, and what is left after the water is removed is all protein. Because their diet lacks fats and carbohydrates, vampire bats are not able to store energy for very long and they can starve to death if they go for only two nights without feeding. A vampire bat drinks about two tablespoons (about thirty milliliters) of blood each night and will sometimes share a blood meal with another vampire bat that did not have a successful night's hunting by regurgitating blood into the hungry bat's mouth. The thin, stretchy walls of the vampire bat's stomach expand after a blood meal, so the bat appears bloated and is sometimes unable to fly without resting for a while. Tiny blood vessels surround the stretchy part of the stomach and facilitate rapid absorption of the blood protein and excretion of large amounts of water in the form of urine.

Common vampire bats have only twenty teeth, fewer than any other species of bat. This is fine since they lap blood rather than chew their food. Their unique feeding habits make the spread of rabies more likely, and the bite wounds they make in cattle increase the likelihood of parasitic infections. Unfortunately, beneficial bat species are often killed when cattle ranchers start fires in caves to try to eliminate vampire bats.

TWO

Bat Bodies

Question 1: How are bats different from birds?

Answer: Bats and birds are vertebrates, and many species have comparable diets of insects or fruit, with plenty of exceptions, of course. Most birds forage during the day, while most bats forage during the night, so they don't really compete directly for food. Although they both fly, they are very different in many ways. Birds have feathers and a beak with no teeth and they lay eggs; bats have fur and teeth and they give birth to live young that they nurse. Birds are in their own class, Aves, while bats are in the class Mammalia, which includes a range of animals, from mice to elephants.

A bird's wing is made up of a long upper arm (humerus), a forearm (ulna and radius), and the wrist and fingers. The bones of the wrist (carpals) and fingers (metacarpals) are fused together, making the bird wing relatively rigid. The top and bottom surface of the bird wing is composed of feathers. The bat wing also consists of an upper arm (humerus) and a forearm (radius and reduced ulna), but the bones of the wrist and fingers are not fused as they are in the bird wing. Like the human hand, the bat hand has four fingers and a thumb. The arms and fingers are covered with a stretchy membrane made up of a double layer of skin. Sandwiched between the layers of skin are blood vessels, nerves, elastic fibers, and muscle strands. The many unfused joints in the bat wing make it much more flexible than a bird wing. This allows a bat to easily change the

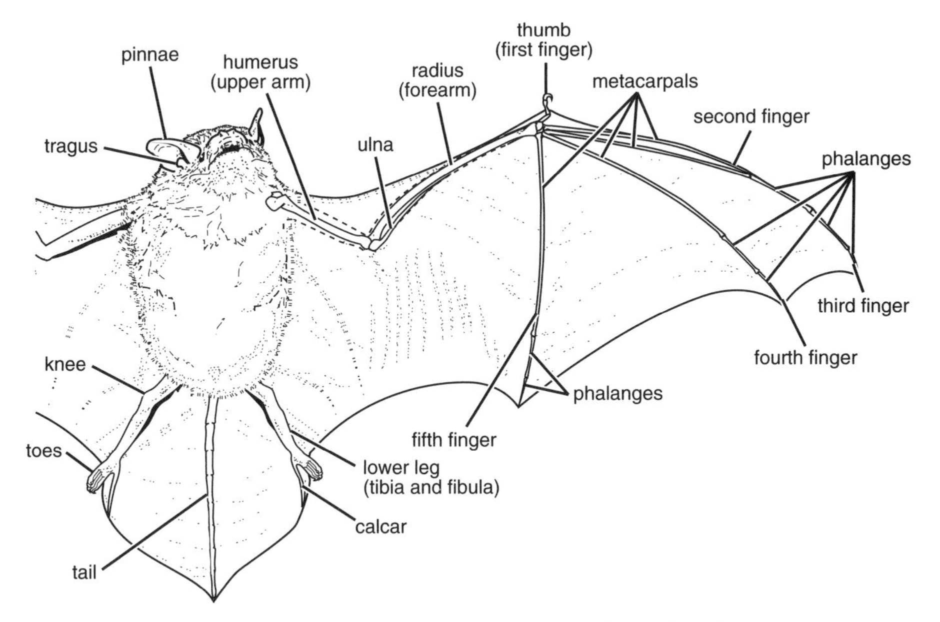

Figure 8. Basic anatomy of a bat. *(Drawing courtesy of David C. Chapman.)*

shape of its stretchy wings so it can change direction rapidly, using one wing independently of the other.

Like birds, many bats also have tails that are useful in flight. Bats have a tail membrane (uropatagium or interfemoral membrane) that extends from one foot to the other, enclosing the tail. The tail membrane in some bat species is supported by a spur of cartilage called a calcar that attaches to the foot. Some bats do not have a tail.

Question 2: Why do bats hang upside down?

Answer: Hanging head down from a roost is an ideal position from which to take flight. Unlike birds, it is difficult for most bats to launch themselves into the air from the ground; they need to drop from a height in order to create enough lift. By hanging upside down, they are in the best position to escape if necessary. When a bat is hibernating, it must use as little energy as possible in order to conserve its body fat, and hanging effortlessly upside down is an excellent way to conserve energy. A bat hanging upside down from the ceiling of a cave or the branch of a tree is out of reach of many predators, and it can easily watch for possible predators below. Microbats have modified cervical vertebrae that make the neck especially flexible so the bat can arch its head backwards, increasing its surveillance area. If a bat detects a predatory snake approaching, it can quickly escape by simply releasing its toes and dropping down into the air to take flight. Hanging upside down by the toes also leaves a bat's wings (hands and arms) free to do other things like manipulate food or hold a pup.

A bat's body is perfectly constructed for an upside-down posture. Its legs are rotated 180 degrees so that its knees and feet point backwards. The bat clings to its roost with the claws on its toes. The weight of the bat's body as it hangs head down pulls on specialized tendons in the toes, keeping them locked in place so that it requires no energy for a bat to hang in this position. When the bat wants to release its grip, it flexes muscles that pull its toes open. Hanging in this way is so effortless that when a bat dies while roosting, it will often remain in place until something knocks it down.

Bat in Different Languages

Abenaki	*madagenihlas*
Algonquin	*nìbàhìgan*
Apache	*ch'abaané*
Arabic	*watwât*, شافخ
Basque	*saguzar, gauenara*
Blackfoot	*mattsiipiitaa*
Bulgarian	прилеп
Caddo	*wakish*
Cantonese	*fûksyú, fèisyú*
Cebuano	*kabog*
Cherokee	*dlameha*

Bat in Different Languages

Cheyenne	*mosheshkanetsenoonahe*
Chinese	*biânfú,* 棒
Chipewyan/Dene	*tsáret'anáaze*
Choctaw	*halambisha*
Cree	*Apahkwaces*
Croatian	*šišmiš*
Czech	*netopýr*
Dakota Sioux	*xupahuwakihdakena*
Danish	*flagermus*
Dutch	*vleermuis, knuppel*
Eklektu	*sots*
English	*bat*
Esperanto	*vesperto*
Estonian	*nahkhiir*
Finnish	*lepakko, yölepakko*
French	*chauve-souris*
German	*fledermaus*
Greek	*nikhteridha,* ρόπαλο
Haida	*k'aats'uu xaalaa*
Hawaiian	*ôpe'ape'a, pe'a, pe'ape'a*
Hebrew	*'atalef*
Hindi	*camgâdar*
Hopi	*sawya*
Indonesian	*kelelawar, kampret*
Italian	*blocco, chirotteri*
Japanese	*Kômori,* バット
Karelian	*uolapakko*
Korean	*pakchwi*
Lakota	*hupahuwakihdakena*
Lappish	*girdisahpan, nahkkesoadji*
Latin	*vespertilio*
Latvian	*sikspârnis*
Lithuanian	*šikšnys*

(continued)

Bat in Different Languages, *continued*

Maliseet-Passamaquoddy	*Motekoniyehs*
Maori	*pekapeka, peka*
Maya	*sootz'*
Mohawk	*yakohonhtariks*
Mojave	*qampanyiq*
Munsee Delaware	*peepiishlongwanaash*
Muskogee Creek	*takfvleleskv*
Nahuatl	*quimichpatlan*
Navajo	*jaa'abaní*
Nez Perce	*'Uuc'uc*
Norsk	*flaggermus*
Norwegian	*flaggermusen*
Ojibway/Chippewa	*Bapakwaanaajiinh*
Olonetsian	*üülepakko*
Oneida	*Tsi'kla'wistal*
Osage	*Tsebthaxe*
Passamaquoddy	*motekoniyehs*
Pima/ Tohono O'odham	*Nanakumal*
Polish	*nietoperze*
Portuguese	*morcego, bastäo*
Potawatomi	*Mishaknekwi*
Quechua	*mashu*
Romanian	*lilieci*
Russian	*letuchaya mysh'*, летучая мышь
Serbo-Croatian	*slepi mi^s*
Shawnee	*pithágathá*
Shoshone	*Honobichi*
Slovakian	*netopierov*
Spanish	*murciélago*
Swahili	*popo*
Swedish	*slagträ*
Swiss German	*flaedermus*
Tagalog	*paniki, kab'g*

Bat in Different Languages

Taiwanese	*li bol*
Thai	*khâang-khâaw*
Tlingit	*S'igeideetan*
Turkish	*yarasa*
Vietnamese	*con do'i*
Welsch	*ystlum, slumyn*
Yaqui	*Sochik*
Yiklamu	*bami*

Native American language translations by Laura Redish with the nonprofit organization Native Languages of the Americas, www.native-languages.org.

Question 3: Do bats have teeth?

Answer: Like other mammals, bats have teeth for chopping their food into smaller pieces so it can be more easily digested. Most bat pups are born with tiny, sharp, hooked "milk" teeth, so-called because they press against a mother's teat as the pup nurses. Within a few weeks as the young bat grows, the milk teeth fall out and are replaced by a full set of adult teeth. All bats have four canines, two on the top and two on the bottom, but the number of incisors, premolars, and molars varies among species.

The common vampire bat (*Desmodus rotundus*) has twenty teeth, less than any other bat. The most teeth are found in some insect-eating bats that can have up to thirty-eight. The surfaces of the teeth in the upper jaw fit perfectly together with the surfaces of the teeth in the lower jaw. These surfaces are flattened in fruit-eating bats so that they can crush fruit, and in insect-eaters they have sharp W-shaped ridges that are suitable for tearing insects apart. After food has been chopped up by the teeth, it moves rapidly through the digestive system, aided by enzyme activity. Some bats have an enzyme called chitinase that

Figure 9. The teeth of an adult Brazilian free-tailed bat (*Tadarida brasiliensis*). (*Photograph courtesy of Bat World Sanctuary, www.batworld.org.*)

digests the external covering on insects which is made of chitin. Chitinase can even be at work in a bat during hibernation. In the case of fruit-eating bats, enzyme activity acts on the seeds in the fruits the bats eat, which helps the seeds to germinate when they fall to the ground in the bats' droppings (see chapter 7, question 3: Are bats useful to plants?)

Question 4: Can bats walk?

Answer: Bats are spectacular fliers, but they can be vulnerable on the ground because they usually need to drop from a perch in order get enough lift to fly. Many species of bats can move well for short distances on the ground, crawling on their feet and wrists, but other species are only able to hop awkwardly. Among the most capable movers, New Zealand short-tailed bats (*Mystacina*) have short, thick legs and talons on the claws of their feet and thumbs, and they are able to run freely on the ground and to climb smooth surfaces. Sucker-footed and disk-winged bats (*Myzopoda* and *Thyroptera*) have disk-shaped suction cups on their feet and at the base of their thumbs that help them cling to the smooth surfaces of stems and palm leaves.

Perhaps the most accomplished walker is the common vampire bat (*Desmodus rotundus*), which has very strong arms and legs and is able to walk and jump well. In an experiment, Daniel Riskin, John Hermanson, and Gerald Carter of Cornell University set up a treadmill to observe vampire bats as they walked; videotapes revealed that the bats tucked their wings under their forearms and used the power of their front limbs, like a gorilla, to run. As the treadmill speed was increased, they were observed running at speeds of almost four feet (close to 1.2 meters) per second. Since the treadmill could not be programmed to move faster, this may not even represent the bats' top speed. The evolution of the vampire bat's superior walking ability remains an interesting puzzle.

Question 5: How fast do bats grow?

Answer: Bats of different species grow at different rates; but, in general, bats that live in temperate regions grow faster than those that live in the tropics. Faster growth rates are essential in temperate regions because young bats must reach maximum growth and have sufficient fat reserves prior to migration or hibernation, when the cold weather sets in.

The amount of food that is available to a lactating mother is one of many factors that influence the growth rate of her young. For example, insect abundance can be affected by temperature and precipitation, and during seasons when insects are less abundant, pups grow more slowly than they do in years when insects are plentiful. Cold temperatures can cause mothers and pups to spend more time in torpor, which also slows growth.

Bat pups are large at birth, weighing from one-fifth to one-quarter of the mother's weight. Many are born naked, although some, including the megabats, are born fully furred. Growth rates vary between species, although most young are not able to fly until their wings have reached 90 percent of the adult wing dimensions. Some young are weaned and flying within three weeks, while others are not ready to fly until the sixth or seventh week after birth; large megabats may not begin flying until they are three months old. The common vampire bat (*Desmodus rotundus*) has the slowest growth rate of all bats studied to date, and their young are not weaned until they are at least seven months old. Although the reason for their slow growth is not known for certain, it could be related to the fact that they feed exclusively on blood, which is deficient in carbohydrates and fats.

Question 6: Are bats blind?

Answer: Someone with poor vision is commonly called "blind as a bat," but the expression is inappropriate since bats can actually see quite well, with visual acuity varying from one species to another. Both megabats and microbats rely on vision during social interactions with one another, to watch for predators, and for navigating across landscapes. Megabats have large eyes and depend on vision to orient themselves during flight and to find food. Most microbats use echolocation to navigate and find food, and they tend to have smaller eyes, although they, too,

Figure 10. *Myotis velifer,* the insect-eating cave myotis, is a microbat that has small eyes. *(Photograph courtesy of Barbara A. Schmidt-French.)*

use vision during their daily activities and to detect objects outside of the effective range of echolocation, which is about thirty-three to sixty-six feet (ten to twenty meters). Some bats are also capable of visual pattern discrimination, which may assist fruit or nectar bats in finding food.

The retinas of most bats consist mainly of rod cells that require very little light to be activated, thus all bats have night vision in low light conditions. While not much color is visible in dim light, the retinas of megabats also contain cones that, when activated by brighter light, enable them to see some color. Until recently, it was thought that only megabats have color vision, but new research suggests that some bats in the microbat family Phyllostomidae are able to make distinctions between red and green.

Question 7: Why do bats have big ears?

Answer: Not all bats have big ears; in fact, the ears of megabats are relatively small and simple. They use their acute hearing to listen for predators and to communicate with one another, but they rely mainly on vision and smell to find the fruit or nectar that they eat. Echolocating microbats also rely on hearing to detect predators and to communicate with one another, but in addition they need to amplify sound when they hunt for insects or other prey using echolocation. Their ears are more prominent and have a variety of tiny folds and notches that help to collect returning echoes and sounds produced by small insects or other prey (see chapter 4, question 1: How does echolocation work?). Microbats that listen for the faint sounds made by prey moving on the ground or on foliage have the largest ears of all bats. With its exceptionally large ears, the African false vampire bat (*Cardioderma cor*) can hear the footsteps of a beetle walking in sand from six feet away (1.2 meters).

The external part of each ear is called *pinna,* and bats can rotate and tilt their pinnae to pick up sounds. Most microbats also have a fleshy projection at the base of each pinna called a *tragus.* The tragus is believed to assist echolocating bats in the

Figure 11. *Euderma maculatum,* the insect-eating spotted bat, has large pink ears for the reception of echolocation calls. *(Photograph courtesy of* M. D. *Tuttle, Bat Conservation International, www.batcon.org.)*

localization of horizontal targets. Bats in the genera *Corynorhinus, Idionycteris,* and *Euderma* are unusual in that they can roll their long ears up on the sides of their heads when they are resting so that they resemble the curled horns on a ram's head.

As in other mammals, the external ears direct sound waves through the ear canal, causing the ear drum (tympanic membrane) to vibrate. The vibrating ear drum then causes the vibration of three small bones (auditory ossicles) directly behind it in the air-filled cavity of the middle ear. The three small bones are named according to their shape, that is, the hammer (malleus), the anvil (incus), and the stirrup (stapes). The movement of these bones causes the vibration of a membrane-covered opening between the middle ear and the inner ear. The vibrating membrane then transfers the sound energy to the cochlea of the inner ear, which is composed of three fluid-filled tubes embedded within a bony capsule. Movement of the fluids in the inner ear stimulates sound-sensitive hair cells that send signals to the brain, where sound perception takes place.

Because some bats emit very loud echolocation calls, they need a way to reduce their own sensitivity to the calls as they make them. This is accomplished by the contraction of two muscles of the inner ear, the tensor tympani and the stapedius. Contracting these muscles decreases the transmission of sound vibrations within the bat's head by causing the stiffening of the small bones in the middle ear, which reduces the intensity of certain frequencies. By contracting these muscles a few milliseconds after producing each echolocation call, the bat is not deafened by its own loud sounds (see chapter 4, question 1: How does echolocation work?).

THREE

Bat Life

Question 1: What do bats eat?

Answer: Bats eat a surprising variety of foods. About 70 percent of bats are insectivorous, meaning they eat insects such as moths, caterpillars, beetles, flies, grasshoppers, planthoppers, leafhoppers, crickets, termites, mosquitoes, and flying ants. A single bat can eat more than one thousand small swarming insects, such as midges (Chironomidae), in an hour. Some insect-eating bats, such as the pallid bat (*Antrozous pallidus*) and species of slit-faced bats in the genus *Nycteris,* also include scorpions in their diet. In Bracken Cave in central Texas, a colony of as many as twenty million Brazilian free-tailed bats (*Tadarida brasiliensis*) emerges in a huge cloud at dusk. Michael Novacek described them as "flying vacuum cleaners . . . with their large, flabby mouths opened wide . . . sweeping through clouds of insects," consuming up to their own body weight in insects during the course of a night. This bat species is a valuable ally to agricultural interests vital to human health. They consume vast quantities of moths that lay eggs that develop into caterpillars, including serious agricultural pests.

Researchers have used DNA analysis to identify some species of moths that are eaten by Brazilian free-tailed bats, including the corn earnworm or tomato fruit worm (*Helicoverpa zea*), the tobacco budworm, (*Heliothis virescens*), the fall armyworm (*Spodoptera Frugiperda*), and the beet armyworm (*Spodoptera Exigua*). The larvae of these pests feed on an amazing variety

Figure 12. *Myotis yumanensis,* a Yuma myotis, chases a moth. *(Photograph courtesy of Michael Durham, www.DurmPhoto.com.)*

of crops and ornamentals, including alfalfa, apple, artichoke, asparagus, avocado, barley, beet, Bermuda grass, broccoli, cabbage, cantaloupe, cauliflower, celery, collard, cotton, corn, cowpea, cucumber, eggplant, flax, grape, lettuce, lima bean, melon, millet, oat, okra, onion, orange, papaya, pea, peach, pear, peanut, pepper, plum, potato, pumpkin, radish, raspberry, rice, ryegrass, safflower, snap bean, sorghum, soybean, spinach, squash, strawberry, sugarbeet, sugarcane, sunflower, sweet potato, timothy, tobacco, tomato, turnip, watermelon, and wheat.

About 80 percent of the diet of the big brown bat (*Eptesicus fuscus*) is often agricultural pest insects. Both the big brown bat and the evening bat (*Nycticeius humeralis*) have heavy jaws and include beetles in their diet. Many big-eared bats feed on moths. Species of *Myotis* bats eat mainly dipterans (flies and midges), small beetles, and small moths. Other bat species are generalists in their feeding behavior, feeding on a variety of insects. Some insects that bats like to eat, such as termites, ants, and caddis flies, are sporadic in occurrence, although numerous when available. Since many insects are agricultural pests, bats provide an enormously beneficial service at no cost to people or the environment.

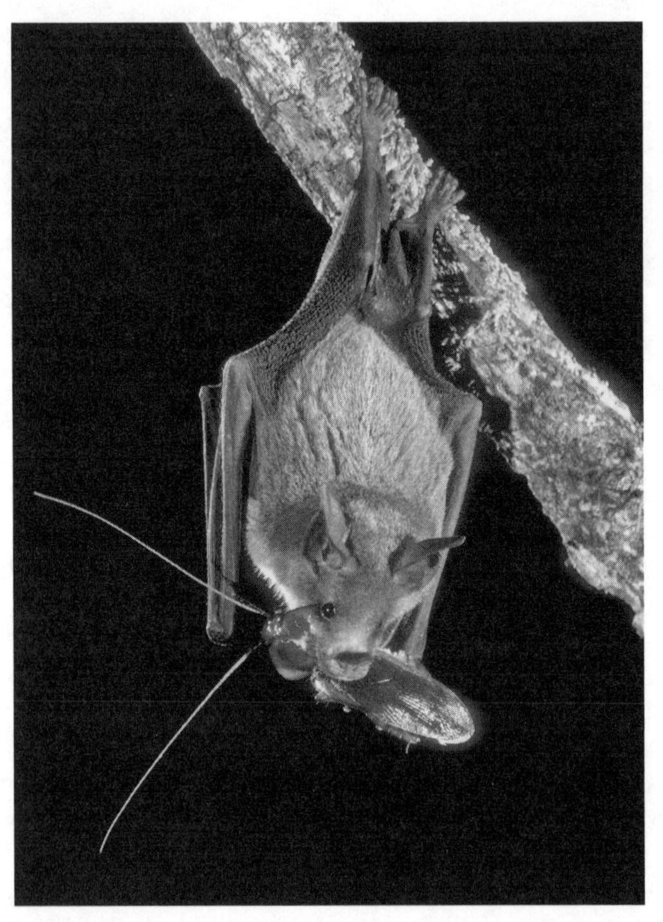

Figure 13. *Micronycteris nicefori,* Niceforo's big-eared bat, captures a roach. *(Photograph courtesy of M. D. Tuttle, Bat Conservation International, www .batcon.org.)*

Most of the remaining species are nectivorous (nectar-drinkers), or they are frugivorous, meaning they are fruit-eaters, sometimes also eating leaves or flowers. Bats pluck fruit from a tree with their mouth, sometimes with the aid of their wings and even their feet. If they carry the fruit away from the tree to eat it, they drop large seeds away from the parent plant, and if they eat fruit with small seeds, the seeds pass through their digestive system, where enzymes help the seeds to germinate when they fall to the ground in the bats' droppings. In these ways, bats help spread the seeds from mangos, peaches, figs, dates, and many other kinds of plants. By distributing seeds over large areas, bats help renew vegetation and aid in the regrowth of rain forests.

Nectar-drinking bats feed on cactus flowers and other plants that bloom at night. Nectar bats have long snouts that fit neatly inside the flowers they prefer, and many have tiny hairs on their tongues that help them lap up the nectar from inside the flowers. While they are drinking, the bats' faces become covered with pollen that they carry to the next flower, helping to fertilize the plant and enabling it to bear fruit. Bats pollinate agave, saguaro and organ pipe cactus, banana, eucalyptus, and many other plants in this way (see color plate D).

Less than 1 percent of all bats feed on small vertebrates in addition to insects. There are about eight species of fish-eating (piscivorous) bats found in the genera *Nycteris, Myotis,* and *Noctilio.* The fisherman bat (*Noctilio leporinus*) has particularly large feet with strong claws that it uses to capture fish swimming just below the surface of the water, which it either eats in flight or carries to a tree to feast on. Piscivorous bats use echolocation to detect ripples on the water from a fish swimming just beneath the surface. The fisherman bat often follows pelicans as they feed, catching the fish that are disturbed as the pelicans dip into the water.

In addition to fish-eating bats, there are a few carnivorous bat species that eat other small vertebrates. In Europe, the giant noctule bat (*Nyctalus lasiopterus*) eats mostly insects but also catches and eats songbirds when the birds migrate annually during the spring and fall. In Latin America, the fringe-lipped bat (*Trachops cirrhosus*) eats small frogs (see color plate A) in addition to insects and lizards. There are also several species of carnivorous "false" vampire bats, including the African false vampire bat (*Cardioderma cor*), which eats beetles, centipedes, scorpions, and sometimes other smaller bats. The carnivorous Australian giant false vampire bat (*Macroderma gigas*) eats mostly mice, but also birds, reptiles, and other bats as well as insects. This bat is known to drop from its perch onto its prey, covering it with the wings and biting the head or neck to kill it. The Asian false vampire bat (*Megaderma lyra*) eats insects, spiders, rodents, birds, frogs, fish, and other bats as well. And there are three species of vampire bats that are sanguivorous, feeding on the blood of birds or mammals (see chapter 1, question 14: Do bats drink blood?). Such dietary diversity among species speaks to the tremendous adaptability of bats around the world.

Like other animals, bats require minerals in their diet, and in caves they may acquire some minerals by licking deposits on cave walls. Christian Voigt discovered that bats in the rain forest in Ecuador visit areas called "salt licks," which are bodies of mineral-rich water and clay. The bats caught in his mist nets at the salt licks were mostly pregnant or lactating females, but outside the salt-lick zones he and his group caught a mix of males and females that were not pregnant or lactating. Voigt suggests that pregnant or lactating females may visit the salt licks to obtain the additional minerals needed to enrich their milk to promote optimal skeletal growth in their pups. This is particularly important because the pups are not weaned until they are almost adult size. The mineral-rich water and clay also detoxify the secondary plant compounds (such as natural pesticides and other toxic chemicals) that the bats take in when they eat fruit, compounds that can be damaging to their embryos or pups. Because minerals have been depleted from the soils and crops in many tropical areas, some indigenous people in South America and Africa also eat mineral-rich clay to supplement their diet.

Bat Families of the World

Megachiroptera
Pteropodidae, Old World fruit bats

Microchiroptera
Craseonycteridae, hog-nosed bats
Emballonuridae, sheath-tailed bats, sac-winged bats, and ghost bats
Furipteridae, thumbless bats and smoky bats
Megadermatidae, false vampire bats and yellow-winged bats
Molossidae, free-tailed bats
Mormoopidae, naked-backed bats and mustached bats
Mystacinidae, New Zealand short-tailed bats
Myzopodidae, disk-winged bats
Natalidae, funnel-eared bats
Noctilionidae, bulldog bats
Nycteridae, slit-faced bats
Phyllostomidae, New World leaf-nosed bats
Rhinolophidae, horseshoe bats
Rhinopomatidae, mouse-tailed bats
Thyropteridae, sucker-footed bats
Vespertilionidae, vesper bats

Question 2: Where do bats live?

Answer: Bats can be found all around the world except in the polar regions. In the vicinity of the Queen Charlotte Islands off the northwest coast of British Columbia, Keen's myotis (*Myotis keenii*) raise their young alone, often in geothermally heated rock cavities. But most species live together in groups called colonies, and species that form large colonies often live and raise their young in caves. Small colonies sometimes roost in cavities in a cave ceiling where heat is trapped. One of the largest colonies ever observed consists of as many as twenty million Brazilian free-tailed bats (*Tadarida brasiliensis*) living and rearing their young in Bracken Cave in central Texas. When the cave is occupied by the bats, their combined body heat can raise the temperature by more than 20 degrees Fahrenheit (6.7 degrees Celsius), providing the warmth needed for the young pups to thrive. Brazilian free-tailed bats are active year round. When winter arrives, those in the southwestern United States migrate south to warmer climates, while those in the southeastern United States remain in the same region throughout the year.

Other bat species hibernate in caves or other secluded places where temperatures are stable during the winter. Big brown bats (*Eptesicus fuscus*) often hibernate near cave entrances, where they cluster together and wedge their bodies into crevices. In the northeastern United States, big brown bats often hibernate in buildings. Tri-colored bats (*Perimyotis subflavus*) roost in somewhat warmer areas of caves than the big brown bats, and they roost alone rather than clustering together like colonial species. Indiana bats (*Myotis sodalis*) pack into dense clusters in a cave. Canyon bats (*Parastrellus hesperus*) hibernate in deep crevices on the sides of cliffs.

Some bats commonly live and raise their young in plants or trees, roosting in leaves and branches, under bark or inside hollow tree trunks. Banana bats (*Musonycteris harrisoni*) live inside furled banana leaves, and when the leaves unfurl as they grow, the bats move into new ones. Disk-winged bats (*Thyroptera* sp.) roost in the curled leaves of plants such as *Heliconia,* using suction cups on their feet and wrists that allow them to hang head-up and to move in and out of the slippery leaves. The Honduran white bats (*Ectophylla alba*) chew along the mid-rib of *Heliconia* leaves, causing the sides to fold down like a tent to shelter them from the elements (see color plate B). Other tent-making bats modify leaves in different ways to make roosts.

Fur color provides camouflage for many bats (see chapter 6, question 7: How do bats avoid predators?). Chocolate brown fur with white tips and pale zigzag lines on the lower back helps the Latin American Proboscis bats (*Rynchonycteris naso*) blend into the bark of mangrove trees where these bats often roost in small groups, heads facing down, forming a vertical line on the tree trunk. Flying foxes (*Pteropus*) live in tropical areas of Asia, Africa, and Oceania and are grayish brown or black, often with yellowish fur between the shoulders. These bats wrap their wings around their bodies and hang from bare tree branches, where they resemble large fruit pods. North American eastern red bats (*Lasiurus borealis*) hang from the stems of leaves or tree branches, where they resemble dead leaves, while yellow bats (*Lasiurus* sp.) are well-camouflaged in dead palm fronds or in Spanish moss that hangs from tree branches.

Shelters made by insects, birds, and other mammals are also popular roosts for some species. In the United States, big brown bats (*Eptesicus fuscus*) in the southwest sometimes roost in old woodpecker holes in desert cacti, and cave myotis (*Myotis velifer*) may roost in abandoned cliff swallow nests. In South America, round-eared bats (*Tonatia silvicola*) are known to roost in cavities in the bottom of termite mounds suspended from tree branches, and in Asia, club-footed bats (*Tyloncteris* sp.) roost in

Figure 14. A hoary bat (*Lasiurus cinereus*) roosts inside a hole in a dead tree. *(Photograph courtesy of Michael Durham, www.DurmPhoto.com.)*

bamboo stems, entering through small holes made by chrysomelid beetles. In Africa, leaf-nosed bats (*Hipposideros fulvum*) sometimes roost in large-crested porcupine burrows, and slit-faced bats (*Nycteris* sp.) have been found living in abandoned aardvark burrows. Perhaps most interesting are Africa's wooly bats (*Kerivoula* sp.), which can be found living in the large webs of colonial spiders.

Due to the loss of natural habitat, some bats now roost in shelters made by people, including buildings, bridges, or abandoned mines. In North America, insect-eating bats often roost in barns or in the attics or crawl spaces of homes. Big (*Eptesicus fuscus*) and little brown bats (*Myotis lucifugus*) commonly roost in buildings in the northeastern United States. Cave myotis (*Myotis velifer*) sometimes roost in road culverts (water pipes), and Rafinesque's big-eared bats (*Corynorhinus rafinesquii*) have been found living in old cisterns or abandoned ammunition bunkers. In Central America, long-legged bats (*Macrophyllum macrophyllum*) have been seen roosting in irrigation tunnels. In Egypt, Tate's trident leaf-nosed bats (*Asellia tridens*) roost in underground channels at oases, and tomb bats (*Taphozous* sp.) sometimes roost in old crypts and pyramids. Brazilian free-tailed bats (*Tadarida Brasiliensis*) readily roost in expansion joints under bridges, and during the summer months, more than a million of these bats live under the Congress Avenue Bridge in the middle of the city of Austin in Texas. In what has become a major tourist attraction, crowds of people assemble every day at dusk during the summer to watch a huge cloud of bats shoot out from under the bridge as darkness settles in.

Question 3: Why do bats like caves?

Answer: Caves provide good homes for bats in many ways. Because caves are underground, the temperatures inside are generally stable and are not influenced significantly by temperature changes that take place above ground. Stable temperatures are necessary for hibernating bats; if temperatures drop too low in a hibernation roost, the bats are at risk of freezing to death. If the temperature becomes unseasonably high, the bats may wake from hibernation too soon, when there are still no insects for them to eat. Stable temperatures are also important for maternity colonies, where mothers raise pups that require warmth in order to grow. Caves are also often inaccessible to animals that might otherwise feed on the bats, and even in caves where predators can enter, bats often roost on the ceiling, high above the ground where they can't be reached. Caves also tend to have high humidity levels, and most bats require high humidity to prevent dehydration or water loss. Some caves contain minerals like calcium that bats can obtain by licking the walls where they roost. But not all bats roost in caves; some bats roost under tree bark or in hollows in trees, some hang in the foliage of trees or other plants, and many live deep in rock crevices (see chapter 3: question 2: Where do bats live?).

Question 4: Do bats only fly at night?

Answer: Bats and birds coexist because, although they eat a lot of the same things, in general they occupy different niches in the environment—birds foraging in the day, bats at night. Most birds are dependent on their vision, so they are at a great disadvantage at night, although there are a few exceptions. Owls, for example, are nocturnal birds that are very nimble flyers with excellent night vision, and owls sometimes prey on bats. It is unclear whether bats were nocturnal in their evolutionary past, but if so, perhaps some bats became diurnal because the absence of predators in their habitat allowed them to forage more freely.

Two almost complete specimens of fossilized bats were recently found in Wyoming, representing a new species (*Onychonycteris finneyi*), the oldest species yet discovered. They have provided many interesting clues about early bats (see chapter 1, question 5: When did bats evolve?), and one might expect this discovery to shed some light on whether early bats were nocturnal or diurnal.

Unfortunately, the upper part of the skulls were crushed and their eye sockets could not be examined for clues as to the nature of their vision, but analysis of their anatomy did indicate that they were insect-eaters, capable of powered flight but not of echolocation.

Most bats fly and feed at night and return to their day roosts by dawn. By flying at night, they avoid daytime predators and take advantage of the ample supply of night-flying insects. Exceptions to this pattern are a few species of megabats that fly and feed during the day. These bats are found only on a small number of islands that do not have the usual daytime predators, such as hawks, making daytime activity relatively safe. It is interesting to note that a species that forages during the day on American Samoa, the Samoan flying fox (*Pteropus samoensis*), is seen mostly at night on the island of Fiji, where there are predatory hawks. This species faces another danger; their numbers are diminishing because they are exported to Guam and Saipan for food (see chapter 7, question 4: Do people eat bats?).

On Christmas Island near Australia, the black-eared flying fox (*Pteropus melantus*) can be seen flying and actively searching for fruit during the day. In the Caroline Islands, northeast of New Guinea, the Chuuk flying fox (*Pteropus insularis*) and the Caroline flying fox (*Pteropus molossinus*) also feed during the day. On the island of Nuie in Polynesia, the Pacific flying fox (*Pteropus tonganus*) does the same. Livingston's flying fox (*Pteropus livingstonii*) is a large and rare "partially" diurnal bat, living in the Comoros Islands near Madagascar. It flies from the late afternoon halfway through the night.

There are some island-dwelling microbats that are also diurnal, including the sheath-tailed bat (*Emballonura sulcata*) in the Caroline Islands and the Azores noctule (*Nyctalus azoreum*) that flies in the forests on the Azores Islands in the mid-Atlantic. Some mainland microbats have been observed flying during the day at high latitudes in the summer, including the whiskered bat (*Myotis mystacinus*) found in the Palaerctic region from Ireland and Morocco to Japan and the Himalayas, Daubenton's bat (*Myotis daubentoni*) found from Ireland to Japan, and the common pipistrelle (*Pipistrellus pipistrellus*) that lives in Europe, Morocco, Algeria, Libya, and southwest Asia. Eastern red bats (*Lasiurus borealis*) sometimes fly during the day in temperate forests in North America, and what might have been a maternity colony of diurnal big free-tailed bats (*Nyctinomops macrotis*) was observed in Chihuahua, Mexico, in 1972.

A few bats are active mainly at night but leave their daytime roosts before sunset, including the Central American greater sac-winged bat (*Saccopteryx bilineata*) and the North American canyon bat (*Parastrellus hesperus*). And finally, migrating bats are also sometimes observed flying during the day, contradicting the misconception that all bats only fly at night.

Question 5: What do bats do during the day?

Answer: A few species of bats fly and feed during the day, but most are active primarily at night. During the day, they groom themselves, licking their fur and combing their claws through it; then using their teeth, they remove whatever dirt or debris collected in their claws. They also spend time resting, sleeping, and, if they are a colonial species (bats that live in groups), socializing with one another. When bats socialize, they utter a variety of squeaks, clicks, chirps, buzzes, and trills (see chapter 4, question 6: How do bats communicate?), often interacting with one another physically. For example, some bats gently touch noses in greeting or butt heads during minor disputes, and some young bats have been observed chasing or hopping toward one another in what looks like play.

During the day in the mating season, many kinds of bats mate in the roost (see chapter 5, question 3: How do bats reproduce?). Males of some species sing courtship songs to attract females, but they bare their teeth, bob their heads, and occasionally swat a rival on the head with a folded wing. Dominant males that guard groups of females, called harems, spend time chasing rival males away. Mothers nurse their young during the day and are on guard for predators that might intrude into their roost and pose a danger to them or to their pups.

Figure 15. *Lasionycteris noctivagans,* a silver-haired bat, combs his toes through his fur and then removes debris collected in his claws with his teeth. *(Photograph courtesy of Michael Durham, www.DurmPhoto.com.)*

Question 6: Do all bats live in groups?

Answer: Many bats live in groups, but some live solitary lives. Flying foxes live together in groups called camps, hanging from the branches of trees, and many microbats roost together in groups called colonies. Some bat colonies include both males and females, although they may separate during a part of the year with females forming maternity colonies where they give birth and raise young, while males roost together in bachelor colonies. In some species where females form maternity colonies, each male roosts alone. Roosts can be occupied by only one species or they may be shared. The availability of roosts in the habitat and the degree of protection offered by the roosts influence their occupancy.

Colony size varies depending on the species and sometimes the season. Some colonies include just a few individuals, and others include dozens, hundreds, thousands, or even millions of bats. Some colonial species cluster together, which means they roost close to one another so their bodies touch, while others roost slightly separated from one another. Indiana bats (*Myotis sodalist*) form groups of less than a hundred under tree bark in the summer, but roost in clusters of thousands in caves during the winter. In the summer, a few tri-colored bats (*Perimyotis subflavus*) can be found roosting together in leaves, or they can be found roosting in groups of up to thirty in buildings. Some species group together with single males roosting with multiple females in harems (see chapter 5, question 3: How do bats reproduce?). As many as three hundred bats of some insect-eating species can squeeze together in one square foot.

Question 7: What is bat guano?

Answer: *Guano* is the term used to refer to the droppings (feces) of birds or bats. Bat guano contains the nutrients phosphorous and nitrogen that make it a good fertilizer. Guano from insect-eating bats contains about 10 percent nitrogen, 3 percent phosphorous, and 1 percent potassium. Nitrogen promotes the rapid growth of plants, phosphorous promotes root growth and flowering, and potassium encourages strong stems. In addition to these nutrients, bat guano also contains many of the trace elements necessary for healthy plants, and gardeners who use bat guano claim that, unlike many chemical fertilizers that leach out of the soil soon after they are applied, bat guano continues to enrich the soil and plants for a longer period of time. The

Brazilian free-tailed bat (*Tadarida brasiliensis*) is known as the "guano bat" due to the large amounts of guano produced in caves where hundreds of thousands or even millions of these bats live. Huge vacuums are used to suck guano out of some of these caves and into trucks during times of the year when the bats are not present. The guano is then packaged and sold in nurseries as fertilizer to grow vegetables, fruits, and ornamental flowers. Bat guano is also a source of nutrition for many organisms, including insects, crustaceans, fungi, and bacteria. In 2005, Danté Fenolio at the University of Oklahoma discovered that an amphibian, the blind grotto salamander (*Eurycea spelaea*) living in a cave in northeast Oklahoma, also feeds on bat guano. The cave is home to a colony of endangered gray bats (*Myotis grisescens*). And when a microbiologist examined guano from one cave used by Brazilian free-tailed bats in Texas, he found that a single ounce (about twenty-eight grams) of guano contained billions of bacteria, including many that are beneficial. Some of the bacteria produce enzymes used to detoxify industrial wastes, make natural insecticides, improve detergents, and convert waste byproducts into alcohol. Layered guano deposits in these caves have been used to monitor environmental pollution and long-term climate change.

Bat guano is also a source of saltpeter (potassium nitrate), and in the 1860s a factory near San Antonio, Texas, used saltpeter made from local bat guano to produce gunpowder. Although people often claim that bat guano is also used in mascara, we found no documentation of this. Guanine is a white shiny substance found in the scales of certain fish and in the guano or organs of seabirds and some mammals, including bats. Although guanine is a component of mascara, it is made from fish scales, not bat guano.

Question 8: Do bats migrate?

Answer: Many kinds of bats migrate or make seasonal journeys, moving from one region to another where conditions are more favorable. In temperate regions, some bats migrate short

Figure 16. Some hoary bats (*Lasiurus cinereus*) travel long distances between their summer and winter roosts. *(Photograph courtesy of Bat World Sanctuary, www.batworld.org.)*

distances in the fall to move from their summer roosts to their winter roosts where they will hibernate during the cold winter months when insects are not available (see chapter 3: question 9: Do bats hibernate?). For example, tri-colored bats (*Perimyotis subflavus*) migrate short distances up to seventy miles (112.7 kilometers) between their summer and winter roosts. When spring arrives and insects are available again, the bats wake from hibernation and fly back to their summer roosts where females will give birth to their pups.

Other bats in temperate regions migrate away from their summer roosts before the cold winter months arrive and fly to warmer regions where they will live during the winter. They, too, return to their summer homes when spring arrives. Migration can vary not only from one species to another, but also within the same species. While bats of a particular species in one region might migrate from summer to winter roosts, some populations of that same species in a different region may remain in the same area all year. A good example is the Brazilian free-tailed bat (*Tadarida brasiliensis*). Much of the population of this species in the southwestern United States migrates to warmer regions in Mexico in the fall, some traveling over six hundred miles (approximately one thousand kilometers). But populations of this species that live along the Pacific coast, where temperatures remain mild and insects are plentiful all year, do not migrate. Although they are not capable of prolonged hibernation, if temperatures drop temporarily, the bats may enter a state of torpor for several days at a time in order to reduce their energy demands (see Chapter 3, Question 9: Do bats hibernate?).

Paul Cryan mapped historical occurrence records and chemically analyzed hair samples, using a process called stable isotope analysis, to determine the seasonal movements of migratory tree-roosting bats. Of particular interest is the fact that males and females of some species migrate to different areas. For example, hoary bats (*Lasiurus cinereus*) in North America spend the winter in Mexico and California and then migrate long distances to more northern parts of the continent in the spring and summer. A female hoary bat captured in southern Arizona had a chemical "signature" in her hair stable isotope rating, an indication that she had molted the previous summer somewhere north of the Canadian border. Cryan and others have found that pregnant females of this species begin migrating first, moving into more easterly areas, while males migrate to mountainous areas of western North America. He also found that males of this species tended to conserve energy by entering torpor during the spring migration period when exposed to temperatures below 77 degrees Fahrenheit (25 degrees Celsius), but pregnant females avoided using torpor. This likely reflects the developmental needs of their offspring, since a young bat grows slowly when its mother's body temperature drops.

Although bats that live in the tropics are not affected by cold winters, some still migrate in order to follow flowering plants or ripening fruit as they occur from region to region. By following the plant cycles, they are always in the best position to find the ripe fruit or nectar they prefer. The distance bats migrate can vary from about fifty miles (eighty kilometers) to more than one thousand miles (over sixteen hundred kilometers). Perhaps even more intriguing than how far bats migrate is how they find their way on such long journeys. We know that bats use sight and spatial memory to navigate long distances. Spatial memory is the ability to retain information about the environment, including the orientation of particular details. Richard Holland and associates at Princeton University in New Jersey found that bats also have a substance called magnetite in sensory cells that enables them to detect the Earth's magnetic fields, which likely assists them during migration (see chapter 4, question 2: How do bats navigate in the dark?).

Question 9: Do bats hibernate?

Answer: Many bats that live in temperate regions hibernate during the cold winter months. Small mammals like bats require more calories per gram of body weight than larger ones to maintain a stable body temperature (thermal homeostasis) because they have a greater surface-area-to-volume ratio than larger animals. One way of reducing their demand for energy is to lower their metabolism by entering a state of relative inactivity called torpor. In torpor, the bat's body temperature drops near that of the ambient (outside) temperature and its respiration and heart rate slow down. The heart rate of a bat in torpor can drop to as low as 40 beats per minute, down from a normal resting heart rate of between 250 to 450 beats per minute. During seasons when bats are active, they can decrease their need for energy by going into torpor during periods of the day or night.

When winter arrives and food is not available for a long period of time, bats need to either migrate to warmer climates or go into hibernation, an inactive phase beyond simple torpor. In hibernation, metabolism is decreased to minimal levels for periods lasting from days to weeks at a time. The heart rate of a bat in hibernation can be as low as ten beats per minute. Bats often hibernate underground in caves or abandoned mines where temperatures remain relatively constant and humidity is high enough to prevent them from becoming dehydrated. High humidity sometimes causes water droplets to form on a hibernating bat's fur, further reducing water loss (see color plate F). A hibernation roost (hibernaculum) is typically a quiet place that is safe from predators. As previously mentioned, some bat species squeeze together in tight clusters to hibernate so that they are better able to stabilize their body temperature against possible environmental changes, while other species hibernate alone or in small groups.

Bats in hibernation live off stored body fat they accumulated as winter approached. How long a bat can stay in hibernation depends on how fast it metabolizes its fat reserves. When the temperature rises in the spring, hibernating bats awaken, their heart rate and respiratory rate increase, and blood circulates through a special kind of fat on their back, called brown fat due to its color. Brown fat has a high concentration of cellular structures (mitochondria) that generate heat; the heat generated in brown fat warms the blood that flows through it, which then warms other parts of the body and increases the bat's core body temperature. It takes about ten to twenty minutes for this process to warm the bat enough for it to wake from hibernation. Hibernating bats regularly arouse during the winter to move to another spot in the roost to find a more comfortable temperature, to find water, to urinate, or to carry out a variety of important bodily functions. Each time a bat arouses from hibernation, it uses up about the equivalent of two to four weeks of hibernating fat stores. If a bat wakes too often, it will use up too much body fat and won't have enough to live on for the rest of the winter. For this reason, it is very important not to disturb hibernating bats, and because some species of bats gather in large numbers (tens or even hundreds of thousands) to hibernate, frequent disturbances to these hibernation roosts can cause the death of many bats.

FOUR

Bat Behavior

Question 1: How does echolocation work?

Answer: Echolocation is a sonar-like method used by many bats and a few other animals to navigate or search for food. Many bats use echolocation to find and catch flying insects, to avoid obstacles in their path, and to locate their roost in a cave or other dark place. Some bats also use echolocation to communicate with one another. When an animal echolocates, it makes a sound that moves out into the environment, strikes an object, and is reflected back as an echo. The reception and interpretation of different aspects of the echo provide the animal with information about the size, shape, texture, and movement of the object.

The echolocation calls of most bats are not audible to us because they are ultrasonic, which means they are above the frequency humans can hear (above twenty kilohertz). The intensity or loudness of bats' calls varies, ranging from 50 to 120 decibels, the equivalent of a loud whisper to the sound at a rock concert. Bats that emit low-intensity calls, like the pallid bat (*Antrozous pallidus*), are referred to as "whispering bats." To avoid being deafened when they emit loud echolocation sounds, bats momentarily disengage the sound-detecting structures in the middle ear. Although components of the sounds used by some bats for social communication are audible to humans, other parts of these calls are ultrasonic. Investigating bats in their native Japan, Jie Ma and colleagues found that mother greater horseshoe bats (*Rhinolophus ferrumequinum*) and their pups use echolocation calls to locate each other as the mother moves into the cave where her pup is roosting. As the mother comes closer to her pup, their calls become synchronized.

The term *echolocation* was coined by Donald Griffin at Harvard University around 1938 when he studied bats as they flew through pairs of vertical wires. He noted their accuracy in avoiding the wires, and using new electronic technology, he recorded the ultrasonic sounds made by the bats as they maneuvered between the wires. Then he put earplugs in the bats' ears and let them fly again, and he found that the deafened bats were no longer able to avoid the wires, demonstrating that sound reception was crucial for navigation in these bats.

An echolocating bat produces a series of short, high-frequency pulses of sound by contracting muscles in its voice box (larynx), and these sounds are sent out through its mouth or nostrils. Bats that emit echolocation calls through the nose have strange-looking folds and fleshy projections on the nose and face that may help to amplify or direct the sound. Each pulse of sound bounces off objects in the environment, returning to the bat's ears as an echo that contains information for the bat. The sound waves of the echo are transmitted from the ear drum to the middle ear and then to the inner ear, where the movement of fluids in the cochlea stimulates sound-sensitive cells that send signals to the brain (see chapter 2, question 7: Why do bats have big ears?). George Pollak demonstrated that neural cells in a structure in the midbrain called the inferior colliculus are particularly sensitive to time differences, and it is the time delay between each call and its returning echo that tells a bat the distance to an object. Each cell is tuned to a very specific frequency, and the sound information is integrated and sent from the inferior colliculus to specialized areas in the auditory cortex where different characteristics of sound are processed. An echo will not reach both of a bat's ears at exactly the same time, and the difference between the time an echo reaches each ear helps the bat to determine the exact location of an object. The tiny folds and projections in a bat's ears also affect reception of the echo, providing additional information

Figure 17. The echolocating pallid bat, *Antrozous pallidus*, sometimes captures prey on the ground. *(Photograph courtesy of Bat World Sanctuary, www.batworld.org.)*

about the location of the target. Neural processing occurs in a fraction of a second after each echo is received, and the result is a steady stream of information for the bat about the size, location, and movement of objects in its environment.

When hunting, the duration of each echolocation call and the time between the pulses of sound is dependent on whether the bat is searching for prey, approaching prey, tracking a targeted prey, or catching the prey. While searching for prey, a bat may emit between four to twelve calls per second, and the pauses between calls are longer than the calls themselves. As soon as it detects prey, the bat increases the number of calls per second, and when chasing prey, it may make forty to fifty calls per second. The fast repetition of calls as a bat attacks its prey is referred to as a "feeding buzz."

Some bats make constant-frequency (CF) calls that do not vary in frequency, while others make frequency-modulated (FM) calls that sweep through a range of frequencies anywhere from ten to more than two hundred kilohertz in just a fraction of a second (see chapter 7, question 12: What is a bat detector?). CF signals are longer calls that are good for detecting prey, especially in a cluttered environment, like finding insects in vegetation. CF calls measure the Doppler shift or fluctuation in frequencies of echoes returning from an insect as its wings oscillate back and forth. Information returned in the echoes of CF calls helps the bat determine how fast the prey is moving and

Figure 18. *Rhinolophus yumanensis*, Dobson's horseshoe bat, has a complex nose leaf surrounding the nostrils. *(Photograph courtesy of M. D. Tuttle, Bat Conservation International, www.batcon.org.)*

allows it to distinguish moving prey from immobile objects. The bat compensates for Doppler shift by lowering the frequency of its calls in response to echoes with elevated frequencies, ensuring that the echo will be at the frequency its ears are best tuned to hear.

FM signals can help determine the distance to the prey and some of the details about the prey, especially in open hunting areas. Some bats produce echolocation calls that have both CF and FM components, and when the calls include multiple frequencies, called *harmonics,* one frequency is usually louder than the others. Harmonics increase the effectiveness of FM calls for finding targets.

Bats encounter many sounds when foraging, including the echolocation calls of other bats. How do bats avoid interference from other bat calls, particularly those of the same species? Erin Gillam from the University of Tennessee at Knoxville experimentally tested free-flying Brazilian free-tailed bats (*Tadarida brasiliensis*) to see how they responded when their echolocation calls were played back to them. She found that they shifted the frequency of their calls away from those being played to them. This shift in frequency likely prevents calls of the same frequency from jamming their own calls as the bats forage.

Question 2: How do bats navigate in the dark?

Answer: Because most bats are nocturnal, they need to be able to find their way in the dark, and most microbats use echolocation to navigate and find prey (see chapter 4, question 1: How does echolocation work?). Paul Racey hypothesized that bats in flight can echolocate with very little effort because a bat in flight breathes in a particular way as it flaps its wings, breathing out on the upstroke and in on the downstroke. Since the bat is expelling air during the upstroke, it takes little or no additional energy to also emit an echolocation pulse at the end of the upstroke, making echolocation an energy-efficient method for navigating.

Echolocation is only effective over a relatively short distance of no more than sixteen to twenty-two yards (about fifteen to twenty meters), so bats also rely on sight, spatial memory, and perhaps other senses to navigate long distances. (Spatial memory is the ability to retain information about an environment, including the location and orientation of particular details.) In 2006, Richard Holland at Princeton University and his colleagues explored the possibility that bats may have another sense, an internal sunset-calibrated magnetic compass, which enables them to use the Earth's magnetic field to help them navigate long distances. In their experiments, big brown bats (*Eptesicus fuscus*) were marked with radio transmitters and released one at a time about twelve miles (twenty kilometers) north of their roost, and they were tracked by a small plane. Some of the bats had been deliberately disoriented before their release, and five control bats were released without any experimental intervention. When they were released, the control bats headed south toward their roost with no problem. Ten other bats had been exposed to a false magnetic field for forty-five minutes before sunset and forty-five minutes afterward. Five had been exposed to a field that was 90 degrees clockwise from magnetic north, and the other five had been exposed to a field shifted 90 degrees counterclockwise. The "clockwise" individuals flew due east when they were released north of their roost, while the "counterclockwise" bats flew due west. Five of the disoriented bats got totally lost and had to be retrieved and the other five eventually found their way home. The researchers concluded that these results strongly suggest the existence of a sunset-calibrated internal magnetic compass as a partial explanation for bats' navigational ability.

In 2008, the same group at the universities of Princeton and Leeds conducted a follow-up series of experiments, again using big brown bats and a similar experimental setup. Again only half of the individuals that had experienced the disorienting magnetic field flew in the wrong direction, and all the other bats flew back to their roost. Clearly there is more to learn. Although the response to magnetic fields was not uniform, these studies demonstrate some impact of magnetic fields on bats' ability to navigate.

Batman

Batman, the fictional hero of Gotham, made his debut in May of 1939 during a period of prewar jitters in the United States. The character, created by cartoonist Bob Kane and writer Bill Finger, appeared for the first time in *Detective Comics* #27 (better known by its later name, *DC Comics*). The story centers around a masked crime-fighting superhero whose alter-ego is that of Bruce Wayne, a millionaire business leader and ladies' man whose parents were murdered by a mugger when Bruce was just a boy. The Caped Crusader, as Wayne's Batman persona is also called, became a crime fighter obsessed with crushing evildoers in an attempt to avenge his parents' murder. As a child sitting at his dying parents' bedside, Bruce vowed to dedicate his life to fighting crime, and seeing a bat fly through an open window gave him the inspiration for his secret character. Bats in the Batman stories have a positive image, and Batman appreciates and imitates their special abilities.

One of the hero's most interesting characteristics is his complete lack of "super powers," special abilities, or otherworldly assets; he is utterly human, relying on sleuthing skills, scientific knowledge, technology, and athleticism to best his foes. He uses his scientific skills to create a vast array of weapons, gadgets, and vehicles that are usually modeled after the shape of a bat. We see him throw his Batarang (a bat-boomerang) at fleeing thugs, drive the Batmobile (a bat-car) in a wild street-chase against crime, and fly through the air using his Batgyro (a single-person flying device). Some of his gadgets are inspired by bat physiology or are designed to use bats for assistance. A few of the gizmos are inspired by bats' ability to echolocate and navigate in the dark, including his night-vision goggles and sonar eyepieces, a sonic Batarang, the sonic pulse globes that explode as they come into contact with enemies, an ultrasound scanner, and, most recently, a

(continued)

three-dimensional sonic triangulator used to pinpoint locations of specific individuals over a large area (featured in the 2008 film, *The Dark Knight*).

Both comic book and film versions of Batman have used bat-inspired gliders or high-tech adaptations of his bat cape to fly him around his home town of Gotham City. Batman can even be seen scaling walls like a fantasy bat, using suction gloves, kneepads, or grappling hooks. He also has actual bats at his command, so he can use them as a distraction, specifically, a colony of bats that live in the Batcave under his home, Wayne Manor. Using a sonic device in his boot or, in some versions, in his belt, he summons the flying bats around his location, creating chaos and panic that allow him to escape unnoticed.

Batman should not be confused with another character from the DC universe—Man-Bat. This character also has a fascination with bats' special abilities, but in this series the image of bats as menacing, evil animals is used to represent the character's dark side. The Man-Bat, also known as Dr. Kirk Langstrom, first appeared in *Detective Comics #400,* published in June of 1970. Langstrom was a scientist specializing in the study of bats, and he developed an extract that was designed to give humans the ability to echolocate. He tested the formula on himself in the hopes of remedying his worsening deafness, and although it worked, it gradually transformed him into a "bat-human" monster. Man-Bat first encounters Batman while trying to steal ingredients to reverse his condition. He later manages to develop a serum that gives him control over his transformations and also allows him to retain his human mind and intelligence while in bat form. He works as a detective and crime-fighter for a period of time, but his stability eventually deteriorates and he clashes with Batman again and again in an ongoing battle of good versus evil. Man-Bat can be seen as symbolizing the struggle of a man with his inner

Batman

animal as well as a character that provides a literal antagonist to Batman, mirroring his own inner conflicts with his rage over his parents' murder.

Batman has appeared in many forms of popular media over the last sixty years, including both serials and feature-length animated films of his adventures. The comic-book version continues to the present day. In the 1940s, Batman appeared as a guest on *The Adventures of Superman* radio show. The first filmed adaptation of his story was a thirteen-part serial that began in 1943, starring Lewis Wilson in the title role. It was followed by *Batman and Robin* in 1949, with Robert Lowery as the Caped Crusader. Adam West became one of the best-known incarnations of the role while starring in the campy *Batman* television series that was on the air from 1966 to 1968. This Batman seemed to have a gadget for any situation, and he even had his own go-go dance, the Batusi, a pun on the name of a dance that was popular at the time, the Watusi. Many celebrities guest-starred as villains during the television show's three-season heyday, including Burgess Meredith, Cesar Romero, Eartha Kitt, Julie Newmar, Art Carney, Shelley Winters, Vincent Price, Liberace, Otto Preminger, Tallulah Bankhead, Eli Wallach, Joan Collins, Ethel Merman, Milton Berle, Ida Lupino, and Zsa Zsa Gabor.

The next round of Batman films began in 1989 with *Batman,* directed by Tim Burton and starring Michael Keaton in the title role. Many consider this to be the definitive film and the most satisfying portrayal of the character. While maintaining a sleek and gothic tone, the story unfolded with sly humor. Both director and star returned with a sequel, *Batman Returns,* in 1992. Two more sequels followed, both directed by Joel Schumacher: *Batman Forever* in 1995, with Val Kilmer as the Dark Knight, and *Batman and Robin* in 1997, starring George Clooney.

The franchise received a complete reboot from director

(continued)

Batman, *continued*

Christopher Nolan in 2005 with *Batman Begins*. He took a much darker tone and, with Christian Bale as his star, explored the sources of this strange hero's motivation, beginning with young Bruce witnessing his parents' brutal deaths. Nolan and Bale returned in 2008 with a sequel, *The Dark Knight,* which received rave reviews from both fans and critics and was a record-setting box-office success.

What comes next? We'll have to wait and see, but obviously the bat-inspired character holds an enduring fascination for a wide audience, and Batman will continue to entertain us.

Aisha C. Butler

Bats and other mammals like dolphins and humans, as well as many birds (including homing pigeons) and some insects (butterflies and honeybees), have a substance in some of their cells called magnetite, a magnetic iron oxide. Richard Holland and his colleagues hypothesized that the disorientation of some of the bats was due to the effect of the magnetic fields on the magnetite in their bodies. Although humans have magnetite in their bodies, they have lost, or perhaps never had, the ability to respond to magnetic fields.

In another study, Yinan Wang at the Chinese Academy of Sciences in Beijing and his colleagues exposed *Nyctalus plancyi* bats in a dark experimental chamber to an altered magnetic field at twice the intensity of the normal magnetic field in that region. The bats were observed under infrared light to simulate nocturnal conditions, and when the magnetic field was experimentally reversed, the bats consistently responded by changing their hanging positions, moving from the northern to the southern end of the chamber. Again, bats were demonstrating an apparent response to magnetic fields, and the researchers speculated that the bats' response to magnetic polarity may help them choose warmer roosts.

Question 3: Do all bats use echolocation to find food?

Answer: Most microbats rely on echolocation for catching their prey in the dark. Insectivorous species can precisely target a flying insect, and fish-eating species can detect a fish just beneath the surface of the water. Vampire bats echolocate to orient during flight, but they seem to rely on smell when they get close to their prey. Since they tend to return to drink blood from the same prey night after night, spatial memory may also be a factor in their behavior. Some species, like the carnivorous false vampire bats, rely more on vision and passive listening to locate prey, hanging quietly from a perch and listening for the sounds of insects or small vertebrates moving around on the ground.

Fruit bats in the Americas (New World) are microbats. They use a combination of echolocation and smell to pinpoint the location of fruit, and some bat-pollinated flowers have developed shapes that are attractive to echolocating bats because they are acoustically conspicuous—that is, their shape yields prominent echoes that the bats can easily discern. In other parts of the world (Old World), bats that eat fruit or nectar are referred to as megabats. Megabats do not echolocate, but they have an excellent sense of smell, large eyes and good vision, and they depend on these senses to find food.

Fruit and nectar bats also use spatial memory, enabling them to return directly to the same group of blooming flowers on subsequent nights as long as the flowers continue to offer nectar. Richard Holland and his colleagues conducted experiments confirming that Egyptian fruit bats (*Rousettus aegyptiacus*) have the ability to recall details of the location of a perch even when local landmarks are moved, and there is an extensive body of research going at least as far back as W. L. Hahn's work in 1908 which confirms the existence of spatial memory in different species of bats. This ability to recall details from their environment minimizes the amount of energy bats need to expend in searching for food.

Question 4: Are bats the only animals that use echolocation?

Answer: There are a few other mammals and some birds that use echolocation to navigate or to find food, including dolphins, porpoises, and killer whales. Water is a particularly good medium for transmitting sound. Dolphins take advantage of this by using echolocation for communication and for locating fish and other objects in the water by sending out a series of high-frequency clicks from nasal air sacs. A fatty organ in the head, called the melon, acts like an acoustic lens to modulate the sounds, and the returning echoes are received by the jaw and transmitted through bodies of fat to the inner ear.

Wandering shrews, short-tailed shrews, and probably many other shrews, as well as the small, mouse-like tenrac of Madagascar, use echolocation to find insects. Although they are fruit-eaters and they forage by sight, nocturnal cave-dwelling oilbirds native to South America navigate by echolocation under poor light conditions by making a series of sharp, audible clicks. Several species of swiftlets from Southeast Asia forage for insects during the day and use audible clicks to echolocate when they are navigating in the dark caves where they roost and breed.

Question 5: How does a bat's prey defend itself?

Answer: While echolocation, also called biosonar, makes bats superb predators, the evolutionary development of defensive strategies helps many small animals and insects avoid capture by bats. Rodents and other small vertebrates on which carnivorous bats feed sometimes escape by using the simple defense of entering a hole or burrow. Insects are the favored prey of many microbats, and even insects that are active during the day sometimes migrate or disperse at twilight or nighttime when bats are active. Some insects can hear an approaching bat's echolocation calls, and this has resulted in an "evolutionary arms race" of strategies and defenses that continue to yield a rich body of research. Many insects have ears or tympanal organs that consist of an external thin membrane covering an air- or fluid-filled sac with sensory cells attached. These organs are located on various parts of the body, such as the wing or abdomen, depending on the species. These tympanal insects produce complex acoustical defense behaviors used for intraspecies communication predating the appearance of bats. Indeed, as Lee Miller and Annemarie Surlykke at the University of Southern Denmark point out, insects existed at least 300 million years ago, giving them a 250-million-year head start before bats appeared.

Some Lepidoptera (moths and nocturnal butterflies of the superfamily Hedyloidea that have ears on their wings) display antibat tactics, as do Orthoptera (crickets), Dictyoptera (praying mantids), Neuroptera (green lacewings), and possibly some Diptera (flies) and Coleoptera (beetles). Their defenses include a variety of evasive techniques: flying away from the source of the ultrasound (negative phonotaxis); power diving, flying toward the ground, or passively falling to the ground when they are startled by the ultrasound (acoustic startle response); flying in spirals or loops; flying into thick vegetation, which returns confusing signals to some echolocating bats; stopping in their tracks if they are walking; and even, in the case of some Pyralid and Noctuid moths, aborting a sexual approach to a pheromone-emitting female.

Insects may be warned of a bat's approach if they can hear ultrasonic calls, but the frequencies they can hear are limited, and some bats adapt by using frequencies that are higher or lower than the hearing range of their prey. This strategy has limitations for the bat, however, because of frequency-distance tradeoffs. Higher frequencies result in higher resolution but cover a shorter distance, and lower-frequency calls have a greater range but lower resolution, making it possible to detect only larger insects. Studies have shown that large moths can be detected by bats from as far away as ten meters (nearly 33 feet), but moths that are able to hear can detect an approaching bat from one hundred meters (328 feet) away, giving them a ten-fold margin of safety. Moths also have other defenses such as evasive flight behavior. Some moths produce a noxious, foul-smelling froth as a defense when they are attacked. Others have chemical defenses in the form of sequestered toxins obtained from the plants in their diet, and they advertise their toxicity with high-frequency clicking sounds.

Some researchers believe that the purpose of the clicks made by tiger moths is to confuse an echolocating bat by making phantom echoes and "jamming" its biosonar, although there have been no free-flying studies of bats and insects to confirm this hypothesis. Nikolay Hristov and William Connor found to the contrary that inexperienced bats of some species were not deterred by the clicks, although after tasting a toxic moth, they quickly learned to associate the clicks with unpalatable prey to be avoided. Jesse Barber and William Connor found that clicking sounds are also used as a defense by some palatable moths that mimic the clicks of the toxic species in order to discourage predators.

Jens Rydell observed male ghost swift moths (*Hepialus humuli*) that are frequently attacked by local bats. Ghost swift moths do not have ultrasonic hearing to warn them about the approach of an echolocating bat, but they have two defenses that minimize their losses. They limit their mating display flights to a thirty-minute period at dusk, and the rest of the time they fly very close to the vegetation, making them hard to locate because of the confusing acoustic background. His research estimated that 20 percent of these moths were eaten each night by bats, but without these strategies their losses would be much greater.

Question 6: How do bats communicate?

Answer: Like many other mammals, bats communicate with one another using physical interactions like nuzzling, butting heads, and swatting one another with folded wings. They also use scent and displays (posturing) to attract a mate or to discourage rivals. But what is most fascinating about some bats is their use of complex vocal communication in a way that seems to resemble speech. Amanda Lollar and Barbara Schmidt-French observed Brazilian free-tailed bats (*Tadarida brasiliensis*) in captivity for over a decade, and they found that the bats appeared to use a variety of calls or vocalizations to locate, greet, argue, and perhaps even play with one another. They also recorded the males singing to females. Schmidt-French, believing these bats may be using a simplistic form of syntax in their communications, began working with George Pollak and his associates at the University of Texas Institute of Neurosciences in Austin. Together with Kirsten Bohn and other colleagues at the Pollak lab, they recorded many of the Brazilian free-tailed bat calls. Bohn's analysis demonstrated that the calls vary not only in the individual notes (called syllables), but also in the sequence and timing between the syllables. This fascinating finding suggests that the bats may combine specific sounds or phrases in different ways to convey specific meanings, reminiscent of human language. They also found that individual males in the captive colony sang slightly different versions of the courtship song and that individuals maintained the same version of that song from year to year. In another interesting observation, Mirjam Knörnschild and colleagues recorded baby sac-winged bats (*Saccopteryx bilineata*) "babbling," uttering long strings of adult-like noises that appear to be precursors of the well-defined social calls they use as adults.

Colonial bats often communicate with one another in their roosts, and scientists have found evidence that bats also communicate while they are in flight. In addition to courtship songs sung by males to attract mates, distress calls and group foraging calls have been documented. Gerald Carter showed that elusive white-winged vampire bats (*Diaemus youngi*) rapidly exchange vocalizations used as "contact calls" for finding and identifying one another. As quickly as one-third of a second after the first bat calls, another bat calls, and then others may join in. They make these calls when isolated from one another, while stalking chickens, and when leaving roosts. Jonathan Balcombe and Gary McCracken from the University of Tennessee in Knoxville showed that Brazilian free-tailed bats (*Tadarida brasiliensis*) use unique (signature) calls, which allow mothers and their babies

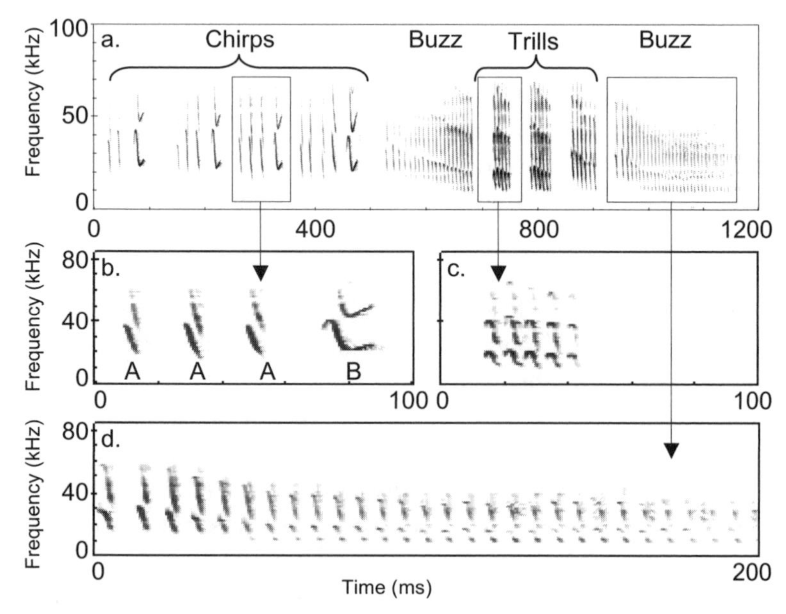

Figure 19. (a) The courtship song of the Brazilian free-tailed bat, *Tadarida brasiliensis,* including three kinds of phrases, a chirp, a buzz, and a trill. (b) Expanded chirp showing two kinds of syllables, syllable A and syllable B. (c) Expanded trill. (d) Expanded buzz. *(Sonogram courtesy of K. M. Bohn, University of Texas, Austin.)*

to find and identify one another. Even solitary bat species communicate with their pups vocally, although their calls may not be unique. Information learned by neuroscientists (like George Pollak) who study bat echolocation and communication in the context of brain functioning may someday be used to better understand how humans develop and process language.

Question 7: How do bats manage extreme heat?

Answer: Like other mammals, bats are endotherms, meaning that they are able to maintain a constant internal body temperature that is regulated by a group of cells in the brain. When active, most bats have a body temperature between 95 and 102 degrees Fahrenheit (35 to 39 degrees Celsius), and they can maintain a desirable body temperature as long as external temperatures do not move beyond critical upper or lower limits. The external temperature range at which a resting bat uses the least amount of oxygen and does not have to draw on stored energy to maintain its body temperature is called its thermoneutral zone.

Since many bat pups are born hairless, clustering together in groups in maternity roosts reduces the amount of energy they require to stay warm. Temperatures in maternity roosts of some colonial bat species may exceed 100 degrees Fahrenheit (38 degrees Celsius). Some bat species keep their young tucked under a wing against their body to protect the infant from ambient temperature fluctuations.

When it gets too hot in a roost, some bats simply move to a cooler roost or to another part of the same roost. If there is still a danger of overheating, the body will respond by increasing blood flow to the wings, which hastens the dissipation of heat from the vascular wing membranes. Flying foxes (Pteropodidae) that hang from the branches of trees flap their wings, lick their fur, and pant to stay cool, but it is still possible for bats to suffer from exposure to excessive temperatures. On January 12, 2002, along the east coast of Australia, Justin Welbergen reported finding at least thirty-five hundred fruit bats, including gray-headed flying foxes (*Pteropus poliocephalus*) and black flying foxes (*Pteropus alecto*), that had fallen out of the trees and died from heat exhaustion. On that day, the temperature rose above 107.5 degrees Fahrenheit (42 degrees Celsius), approximately 56 degrees Fahrenheit (14 degrees Celsius) higher than normal.

Question 8: How do bats manage extreme cold?

Answer: Moving to a warmer roost or a warmer area within the roost is a simple way of responding to low temperatures. Bats can also generate heat by shivering, but doing so requires a lot of energy, and small bats have limited energy resources. As an alternative, they can enter into torpor for periods lasting from days to weeks. Torpor is a state of reduced metabolic activity during which heart rate, respiration, and body temperature decrease, resulting in a reduced need for energy.

Some species of bats native to temperate regions where there are extended periods of cold temperatures migrate to warmer climates where they live during the winter. Other species hibernate in their year-round habitat during the winter, and still others migrate short distances and hibernate in a different location (see chapter 3, question 9: Do bats hibernate? and chapter 3, question 8: Do bats migrate?). Bats gain weight in the late summer or early fall as they accumulate fat needed for energy to

Figure 20. *Myotis evotis,* a western long-eared myotis, drinks from an open pool of water while in flight. *(Photograph courtesy of Michael Durham, www .DurmPhoto.com.)*

survive the cold winter. During hibernation, metabolic activity decreases for longer periods than it does during daily torpor, decreasing to minimal levels; the body temperature drops, heart rate decreases, breathing slows, and oxygen consumption is reduced. Their thick fur coats and increased body fat help to reduce the amount of heat that escapes from the body during this period. Arousing from hibernation requires a significant amount of energy, and a bat can consume two to four weeks worth of fat reserves during one waking incident. If fat reserves are used up too early, the bat may not survive the winter, which highlights the importance of not disturbing bats during hibernation.

Question 9: Can bats swim?

Answer: Most bats are able to swim short distances when necessary, and some bat species that eat fish, like the fisherman bat (*Noctilio leporinus*), are excellent swimmers, using their wings as oars to move through the water. Many bats drink from lakes, ponds, rivers, or creeks while still in flight, swooping down to lap water with their tongue as they fly along. If they accidentally fall into the water, they are usually able to swim to the shore, although they may need to climb out of the water and up a tree so they can drop from a limb to get enough lift to resume flight. Bats that fall into swimming pools while drinking in flight often drown, not because they are unable to swim to the edge, but because they can't get a grip on the smooth surface of the pool to crawl back out.

FIVE

Bat Love

Question 1: How does a bat attract a mate?

Answer: Groups in which multiple males roost with multiple females are common among bats. Males of some of these species establish mating territories either within or away from the main roost. They position themselves in these roosts to attract mates by emitting courtship vocalizations, performing courtship displays by flapping their wings, and marking females and/or territories with their glandular secretions. Groups of male hammer-headed fruit bats (*Hypsignathus monstrosus*) gather at dusk during the breeding season in leks (from the Swedish *leka,* meaning play), for the purpose of courtship display and competitive mating. The large males fly to tree branches along the edge of the group; each bat hangs alone, emitting loud, repetitive calls to attract females, increasing the calling rate as a female approaches. A female will hover near the male, often leaving and returning several times before eventually stopping to mate with him.

At dusk during mating season, male Wahlberg's epauletted fruit bats (*Epomophorus wahlbergi*) emit mating calls in flight as they leave their daytime roosts among the palm leaves. They continue to call in the evening from new roosts, and when a female responds and hovers in front of a calling male, he displays epaulets of stiff yellowish tufts of hair on his shoulders. Male African false vampire bats (*Cardioderma cor*) also leave their communal roosts and fly to new roosts at night, where they establish feeding areas and sing to attract females.

In some species, multiple matings have been observed for both males and females without the establishment of specific territories. Mating has been observed even during periods of hibernation in some species, when males that awaken appear to mate at random with hibernating females.

Question 2: Are bats monogamous?

Answer: Different species of bats form different kinds of social groups for mating purposes, and there can be many reproductive groups of the same type within a colony. A reproductive group of bats can include a single male with multiple females, multiple males with multiple females, or a single male with a single female. Many species of bats form polygamous groups called harems, consisting of a male roosting with more than one female. Some tropical species form seasonal harems, consisting of single males roosting with multiple females to mate during certain times of the year, but many tropical bats form year-round harems. In some species, the group of females in the harem remains together for many years and sometimes even for their entire lives, while other species form harems that are less stable, with females moving back and forth among groups. A harem male defends his group of females and aggressively chases away other males.

Polygamous groups are also found in temperate regions, where males and females hibernate together in the winter and then segregate at or prior to the birth of their pups. In some species, females roost together in maternity colonies where they give birth and care for their young, while males roost separately during this time or roost together in bachelor colonies. Males of some of these species establish seasonal mating territories that are visited by females. The females visit more than one male's territory, mating with several males. *Polyandry* is the term used to refer to a female mating with more than one male.

Of the more than eleven hundred bat species, only seventeen are known to form pairs, with a single male and a single female that are monogamous to some degree. African yellow-winged bats (*Lavia frons*) form pairs that stay together in established feeding territories for several months.

Question 3: How do bats reproduce?

Answer: The basic mechanics of reproduction in bats is similar to that of other mammals. Sperm must be transferred from the male's penis to the female's reproductive tract to fertilize an egg. The egg is eventually implanted in the lining of the uterus, where it develops into a baby bat called a *pup*. In most species, the female gives birth to one pup each year, although there are some species that regularly have twins or multiple births (see chapter 5, question 4: How many pups are in a litter?).

Bats use a variety of strategies for adjusting the timing and length of pregnancy. In temperate regions where insects are more plentiful during warm months, it is common for bats to mate in the fall and then hibernate in the winter. Delayed fertilization occurs in most temperate-zone species; the sperm is stored in the female's uterus until spring when the bats wake from hibernation, and then fertilization and implantation take place. Delayed implantation occurs in a few species; mating and fertilization take place in the fall, but the fertilized egg is stored until the bat wakes from hibernation, and then the embryo is implanted in the wall of the uterus and development continues. Another strategy used by some species is delayed development, in which mating takes place in the fall followed by fertilization, implantation, and early development of the embryo. Then the growth of the embryo stops or is significantly slowed for several months, and normal growth resumes in the spring so that the offspring are born when food is most plentiful. Some species mate in early spring and the pups are born in early summer. In the tropics, reproduction is affected by seasonal rainfalls. Food is most plentiful in the tropics during periods of peak rainfall, and pups of many species are born during the rainy seasons without significant developmental delays.

The gestation period or length of pregnancy varies by species, ranging from about three weeks to six months. Microbat pups

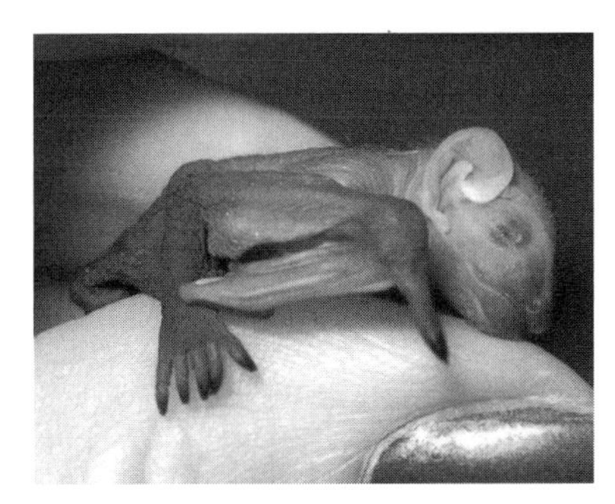

Figure 21. A tiny newborn red bat (*Lasiurus borealis*) sleeps on the thumb of its caretaker. *(Photograph courtesy of Carol Bunyard.)*

are born feet first and can weigh one-quarter to one-third of the mother's weight when born. Megabats are born head first and weigh about one-fifth the mother's weight at birth. Most microbats are only nursed for three to six weeks, at which point they are relatively independent; megabats nurse their young for up to four months.

Thomas Kunz from Boston University reported a fascinating observation of a young captive-bred Rodrigues flying fox (*Pteropus rodricensis*) at the Lubee Bat Conservancy that was struggling to give birth. She was hanging upside down, even though normally bats turn so that they are in a head-up position to give birth. Another female approached and roosted directly in front of the struggling bat, turned head-up, demonstrating the appropriate position for giving birth, and began emulating straining and contractions. Finally, the struggling female turned into the correct head-up position and began giving birth. Unfortunately, the pup was in a breech position, so the "midwife" bat began alternately licking the pup to stimulate birth and fanning the exhausted mother with her wings. The pup was eventually born, exhausted but healthy, likely due to the diligent tutoring of the midwife bat.

Testes size is an interesting area of bat research, considering that in most mammals, larger species have progressively smaller testes relative to body size. But the size of testes among bats has an extraordinary range which is unrelated to the size of the species. The range of testes size in bats is greater than in any other mammalian order and is related to the amount of sperm produced. Gerald Wilkinson (University of Maryland at College Park) and Gary McCracken (University of Tennessee at Knoxville) found that testes are largest in bat species where multiple males roost with multiple females, regardless of the bat's body size (see chapter 5, question 2: Are bats monogamous?). In species where a single male typically roosts with several females, testes size is intermediate; and in species where single males roost with single females, testes size is smallest. In species where the female stores the sperm for several months after mating, the males' testes are relatively large, presumably because they must produce a large quantity of sperm so that enough viable sperm survive the storage period. In another interesting study done by Scott Pitnick at Syracuse University, researchers discovered that in species where the females are promiscuous, the males have evolved relatively large testes but have relatively small brains. Since brain tissue and sperm cells each require significant metabolic energy to produce and maintain, the different species presumably have developed one organ over the other based on which one will improve their chances of reproductive success.

Question 4: How many pups are in a litter?

Answer: Although some tropical bat species give birth two or even three times each year, females of most species give birth to a single pup each year. Big brown bats (*Eptesicus fuscus*) that live in the Rocky Mountains and westward give birth to one pup, while those that live east of the Rocky Mountains commonly give birth to twins, and twins are not unusual in several other species as well. More unique are bats that commonly give birth to larger litters, like the eastern red bats (*Lasiurus borealis*) that commonly give birth to four or five pups in an annual litter. Females that have one or two pups have two teats (nipples) to nurse their young, while bats that give birth to more than two pups, like red bats, have four teats. One pup in a litter of five will not survive because the mother does not have enough teats for all five.

Question 5: Do bat mothers take care of their young?

Answer: All bat mothers nurse their young and provide their pups with warmth, sensory stimulation, and, in some cases, transportation. Bat mothers choose environments that have warm temperatures necessary for their young to grow. Female common vampire bats (*Desmodus rotundus*) begin to regurgitate blood to feed their young when they are about three months old until they are weaned at about seven months. Many bat mothers will retrieve youngsters that fall from roosts and sometimes carry their pups back and forth between different daytime roosts.

Although most bats leave their young in the roost while they fly out to forage at night, mothers of some species actually keep the infant attached to them continuously for the first few weeks of life. Safely tucked under her wing, the pup nurses and shares the mother's body heat. An uncommon condition is found in the African false vampire bat (*Cardioderma cor*); mothers have an extra pair of teats near the genital area called false nipples or holdfast nipples that are not used for nursing, but rather function to give a pup a secure grip during flight. The pup clings to the holdfast nipple with its mouth and wraps its legs around the mother's neck as she flies.

Mothers and young of many colonial species roost together in a group during the day, although there are some exceptions. Mother free-tailed bats (*Tadarida brasiliensis*) cluster together in the same roost as their young, but the young are clustered together with one another. The mothers move to the cluster of young to nurse their own pup at least a couple of times a day. The pups roost together in tight clusters with up to four hundred to five hundred bats per square foot in some colonies.

Bat mothers use multiple senses to interact with their young, including hearing, smell, touch, vision, and spatial memory. Although communal nursing does occur in some species, most females nurse their own infants, so they must be able to recognize and distinguish them from others in the roost. Some use spatial memory to find their own pups in the roost, and some call for their young and listen for their pup's return calls and then

Figure 22. A maternity colony of Allen's lappet-browed bats (*Idionycteris phyllotis*) roosts under loose bark on a dead ponderosa pine tree. *(Photograph courtesy of Michael Durham, www.DurmPhoto.com.)*

use smell to verify the pup's identity. Bat pups often emit low-frequency calls, and mothers are sensitive to these "isolation" calls that are different from the high-frequency calls more commonly used by adult bats. Mothers may have scent glands that they rub against their pups to help identify them, and both mothers and pups probably use visual cues during their interactions. Mothers of some species groom their young, and the young of some species require tactile stimulation from the mother in order to urinate and defecate.

In some species, including the normally solitary eastern red bat (*Lasiurus borealis*), young bats have been observed flying independently along with their mothers when they forage, perhaps learning foraging skills from the mother. John Whitaker at Indiana State University found young red bats with both milk and insect parts in their stomachs, which shows that some mothers were still nursing young that had begun catching insects. Young that still have access to milk when they first start foraging probably have an increased chance of survival. Captive studies of hand-raised red bat pups, conducted by co-author Barbara Schmidt-French and John Whitaker, demonstrated that young were able to learn to capture flying insects and to choose those that were appropriate for their species even without the benefit of a mother. However, it took some of the young a longer period of time to become proficient at catching insects than they would have had in the wild. A mother red bat only has milk available for her young for about a month.

Question 6: Do bat fathers take care of their offspring?

Answer: In most colonial bat species that live in temperate regions, females form maternity colonies where they give birth and raise their young, while males live in other roosts and do not participate in parental care. Males of a few monogamous species do participate in the care of young, including the African yellow-winged bat (*Lavia frons*), the African false vampire bat (*Cardioderma cor*), and Linnaeus's false vampire bat (*Vampyrum spectrum*). Also, males of some species in the tropics defend harem females, including those with pups.

Most unusual are Dyak bats (*Dyacopterus spadiceus*) in Malaysia, Sumatra, Borneo, and Luzon and the Bismark masked flying foxes (*Pteropus capistratus*) in Papau New Guinea. Some adult males of these species secrete small amounts of milk from their mammary glands during the season when young are born. The mammary tissue in these lactating males is the same as it is in lactating females, although the males secrete much less milk. It is not known if these males actually nurse young or not, although it appears they are capable of doing so.

Question 7: How long does it take before newborn bats can fly?

Answer: Young bats must be able to fly by the time they are weaned, and they will need well-developed, adult-sized wings in order to fly and search for food. Most other mammals wean their young by the time they have reached about 40 percent of the weight of an adult, but many bats are not weaned until they are nearly adult size; until then, their wings are not fully developed. For most microbats, this means they are weaned when they are three to six weeks old. In contrast, megabats are able to fly when they are less than adult size, but mothers may continue to nurse their young for several more weeks after they are capable of flight. Some megabats are not weaned until they are about four months old.

SIX

Dangers and Defenses

Question 1: Are bats aggressive?

Answer: Bats are generally shy, reclusive animals that prefer to hide. They are unlikely to attack people, except under unusual circumstances (see chapter 6, question 2: Do bats bite people?). However, some species, like the solitary eastern red bats (*Lasiurus borealis*), chase intruding bats out of their feeding territories while they are flying, and it is not uncommon for these bats to seriously injure one another if they are caught and caged together. Many species live in colonies where large numbers of individuals cohabitate. Bats in these groups often squeak and buzz, sometimes butting their heads together or briskly pushing their muzzles into one another as they squabble and jostle for roosting positions. Most of these interactions do not result in injury.

Bat behavior can change during the mating season, when males of some species are likely to become more aggressive. For example, Brazilian free-tailed bats (*Tadarida brasiliensis*) are generally very mild mannered, but during mating season males in captivity sometimes bare their teeth and bob their heads at one another when protecting mating territories. Occasionally, these disputes become increasingly aggressive, with males swatting one another with folded wings or even biting until one chases the other away. Two captive males may even lock jaws and fall to the floor of their cage, rolling around as they struggle with one another, although this degree of aggression is less common. In the wild, dominant males with harems, like the straw-colored flying fox (*Eidolon helvum*), often aggressively chase off other males that attempt to intrude, squawking loudly and swatting or biting the intruder.

Question 2: Do bats bite people?

Answer: Bats rarely bite people and they are normally not aggressive; their instinct is to hide or to fly away if a person approaches them in the wild. However, if they are touched or picked up, they will often bite to protect themselves, like any other wild animal. Also, a bat with rabies may behave abnormally and fly erratically, even during the day (see chapter 6, question 3: Do all bats have rabies?). About one-half of 1 percent of bats get rabies, and if a sick bat makes accidental contact with a person as it flounders around, it may instinctively bite in self-defense, especially if someone tries to touch it. In order to get up off the ground, a sick or injured bat may climb up onto yard furniture or some other atypical place where, again, it is more likely to have accidental contact with people than would normally occur.

In some parts of the American tropics, people who sleep outdoors or in buildings without screened windows are sometimes bitten by vampire bats, but this is only likely to happen if the bats' normal sources of food are gone. For example, if the bats had been feeding on a herd of cattle and the cattle were moved away, the bats might resort to feeding on people who are easily accessible. People who are bitten by vampire bats are not generally bitten on the neck, as portrayed in movies; it is more likely a bat will bite an exposed big toe, where there are many blood vessels just beneath the skin. Bat bites can easily be avoided by educating children and adults not to touch bats, and in parts of the tropics where vampire bats live, bites can be avoided by sleeping indoors and screening the windows in sleeping quarters.

Question 3: Do all bats have rabies?

Answer: Rabies is a disease caused by a virus in the genus *Lyssavirus,* family Rhabdoviridae. There are seven different strains

A

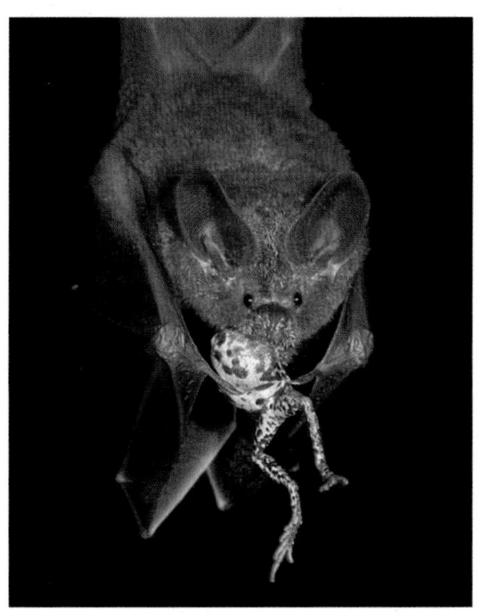

Trachops cirrhosus: the fringed-lipped bat feasts on a Tungara frog. *(Photo by A. T. Baugh, University of Texas, Austin.)*

Megaderma lyra: the Asian false vampire bat includes lizards in its diet. *(Photo by M. D. Tuttle, Bat Conservation International, www.batcon.org.)*

Vampyrum spectrum: Linnaeus's false vampire bat eats prey such as small birds and rodents. *(Photo by M. D. Tuttle, Bat Conservation International, www.batcon.org.)*

Ectophylla alba: white Honduran tent-making bats chew along the midrib of a heliconia leaf, causing the sides to droop down and form a tent to protect them from the sun and rain and keep them hidden from predators. *(Photo by M. D. Tuttle, Bat Conservation International, www.batcon.org.)*

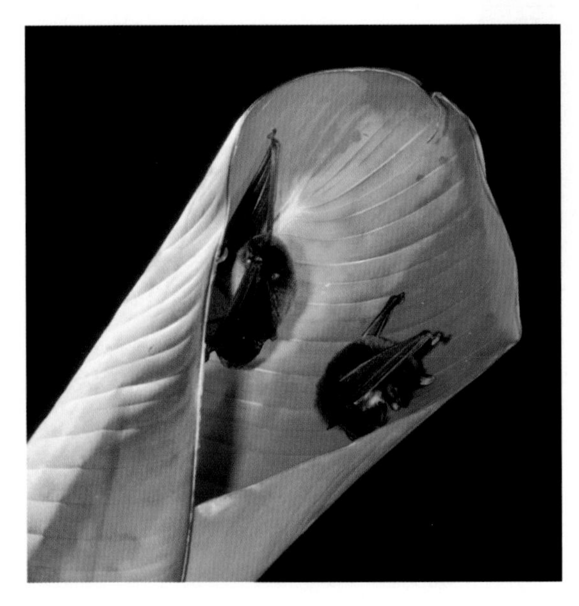

Thyroptera tricolor: Spix's disk-winged bats have suction cups on their feet and the base of their thumbs. Unlike most bats, they roost in a heads-up position. *(Photo by M. D. Tuttle, Bat Conservation International, www.batcon.org.)*

C

Rousettus aegyptiacus: an Egyptian fruit bat enjoys a mango. *(Photo by M. D. Tuttle, Bat Conservation International, www.batcon.org)*

Leptonycteris yerbabuenae: a lesser long-nosed bat drinks nectar from a cardon cactus flower. *(Photo by M. D. Tuttle, Bat Conservation International, www .batcon.org.)*

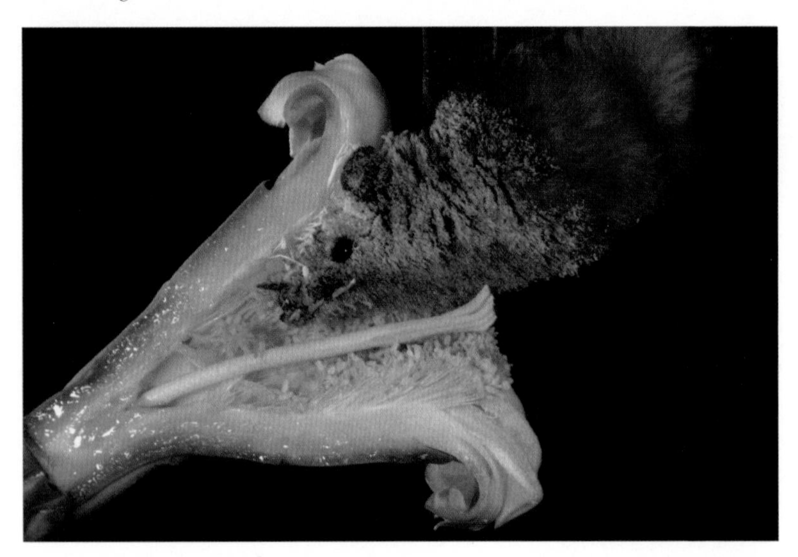

Leptonycteris yerbabuenae: a lesser long-nosed bat pollinates a saguaro cactus (cross section). *(Photo by M. D. Tuttle, Bat Conservation International, www .batcon.org.)*

Dyphylla ecaudata: a hairy-legged vampire bat feeds on the blood of a chicken. *(Photo by G. G. Carter, Cornell University.)*

Desmodus rotundus: two young vampire bats cuddle together in a cave. *(Photo by G. G. Carter, Cornell University.)*

F

Vampyrum spectrum: Linnaeus's false vampire bat lives in parts of Mexico and Central and South America. *(Photo by M. D. Tuttle, Bat Conservation International, www.batcon.org.)*

Myotis mystacinus: water droplets form on the fur of a Eurasian whiskered bat during hibernation. *(Photo by M. D. Tuttle, Bat Conservation International, www.batcon.org.)*

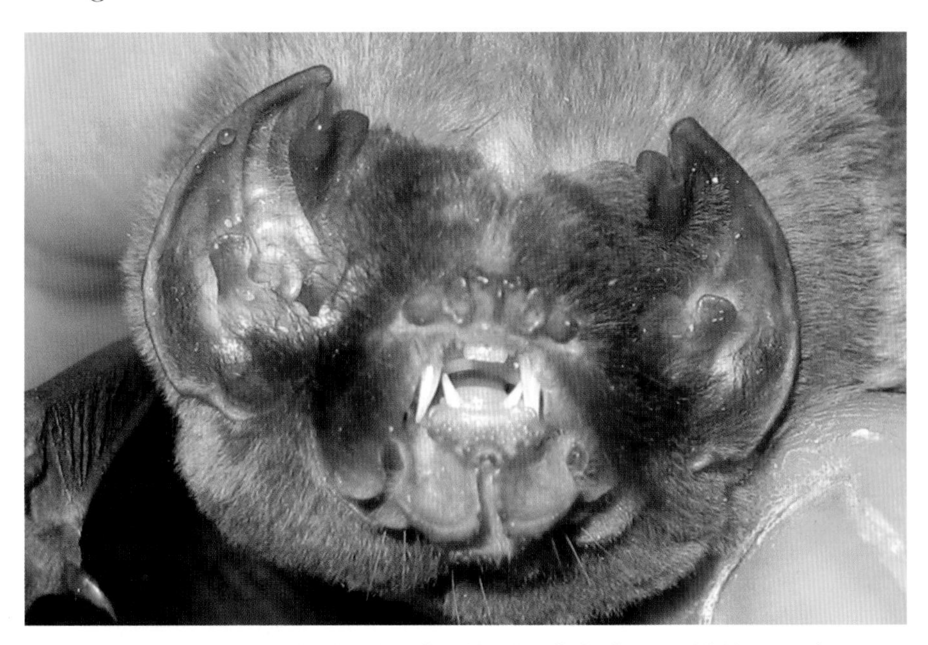

Mormoops megalophylla: the ghost-faced bat has fleshy flaps and folds on its face that may help to direct its echolocation calls. *(Photo by G. G. Carter, Cornell University.)*

Centurio senex: this bat pulls wrinkled skin from its face and chin over its eyes when it sleeps. *(Photo by M. D. Tuttle, Bat Conservation International, www.batcon.org.)*

Artibeus jamacensis: a Jamaican fruit bat is carrying its baby as it flies to another roost. *(Photo by M. D. Tuttle, Bat Conservation International, www.batcon.org.)*

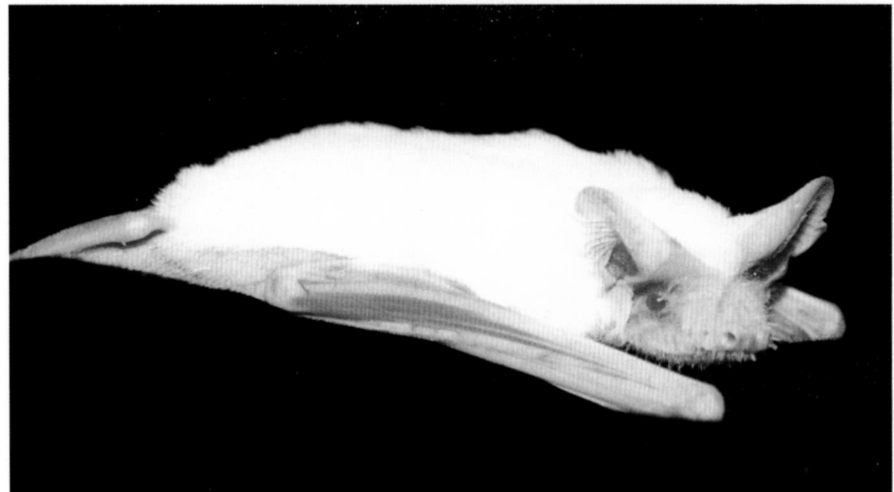

Tadarida brasiliensis: an extremely rare albino Brazilian free-tailed bat roosts in the Bat World Sanctuary in Mineral Wells, Texas. *(Photo courtesy of Bat World Sanctuary, www.batworld.org.)*

(genotypes) of *Lyssavirus,* including classical rabies virus, found worldwide; Lagos bat virus, Mokala virus, and Duvenhage virus in Africa; European bat lyssaviruses 1 and 2; and an Australian bat lyssavirus. Classical rabies is the genotype found in the United States and discussed in this section. It is also the genotype responsible for the majority of human rabies cases worldwide. The vast majority of bats do not have rabies. Although it varies from species to species, surveys by bat researchers have found that only about one-half of 1 percent of all bats have rabies, and that percentage appears to have remained stable over a period of several decades. The disease is transmitted from one infected (rabid) animal to another through a bite, or by saliva making contact with mucous membranes in the eyes, nose, or mouth, or by saliva getting into cuts or abrasions on the skin. Once the virus enters the body, it moves to the central nervous system and causes encephalomyelitis, which is an inflammation of the brain and spinal cord. More than thirty thousand people in the world get rabies from rabid animals every year, and although there are some rare exceptions, it is usually fatal. In Asia, Africa, and Latin America, dogs are the most common animals to have rabies and the most common cause of rabies in people. Human rabies cases are rare in the United States, where there are only a few cases at the most each year. Because many dogs and cats are vaccinated against rabies in this country, most rabies cases are found in wildlife such as skunks, raccoons, coyotes, foxes, and bats.

An extremely rare method of rabies transmission is through the inhalation of viral particles. This route of transmission is believed to have occurred only four times in the United States. Two of the cases did not involve bats and occurred in laboratories where researchers were working with live rabies virus. The other two cases of hypothesized aerosol transmission occurred over fifty years ago in a very unique cave environment where tens of thousands of bats roosted. Although the rabies virus cannot penetrate intact skin, cave gases can cause lesions in the respiratory tract where viral particles could enter. This possible source of transmission is easily avoided by not entering these caves. There have also been a few cases of rabies being transmitted to people through organ transplants.

Vaccinating dogs and cats can help stop the spread of rabies, and some health officials believe the distribution of a bait containing the rabies vaccine may help stop the spread of rabies in some terrestrial wildlife, although some experts dispute the degree to which it has an effect on the population. People can easily avoid rabies by simply not touching wild animals and by seeking immediate medical attention if they are bitten by a wild animal. Medical professionals administer rabies post-exposure prophylaxis, in the form of a series of injections, to people who have been exposed to a rabid animal. The shots include an injection of an immune globulin and five vaccinations that are no longer the painful injections that used to be given in the stomach. Today, rabies vaccinations are given in the arm and are no more painful than any other vaccination, although the vaccine is quite expensive. The series of shots will prevent the development of rabies if they are commenced immediately after an exposure occurs. Some laboratory personnel, bat researchers, rehabilitators, educators, and others who handle bats receive a series of three pre-exposure rabies vaccinations. This decreases the numbers of vaccinations that will be required if they are exposed to a rabid bat. (Rabies vaccinations may not be effective against African *Lyssavirus* genotypes.)

As a note of interest, bats at Bat World Sanctuary are vaccinated against rabies, as are captive bats cared for by co-author Barbara Schmidt-French. Amy Turmelle and associates at the University of Tennessee, Knoxville, and Boston University drew blood from captive Brazilian free-tailed bats (*Tadarida brasiliensis*) at the Bat World Sanctuary and Schmidt-French facility both before and ten days after vaccination. Working with the rabies section of the U.S. Centers for Disease Control and Prevention (CDC), they determined that all of the bats had developed anti-rabies antibodies. These results suggest that administering annual rabies vaccinations to captive bat colonies may minimize the likelihood of rabies infection and transmission between bats.

Question 4: Can people get diseases from bats?

Answer: Bats, like all animals, are susceptible to a variety of parasitic, bacterial, fungal, and viral infections, however they are among the least likely to transmit diseases to humans. In fact, our own pets are far more common disease sources for humans. Rabies is caused by a viral infection in mammals (see chapter 6, question 3: Do all bats have rabies?). Worldwide, most rabies transmission is from dogs. A fungal disease known as histoplasmosis is most often contracted from birds. Infection results from inhaling spore of *Histoplasma capsulatum,* a naturally occurring soil fungus. The fungus is typically found in warm climates in soil enriched by animal droppings. It can be contracted from bat roosts, especially in caves, but seldom causes serious human illness unless large quantities of spore-laden dust are inhaled.

Some scientists have hypothesized that bats may serve as hosts to viruses that cause rare diseases like SARS (Severe Acute Respiratory Syndrome) in China, Nipah (the Malaysian pig virus), Hendra in Australia, Ebola and Marburg hemorrhagic fever in Africa. The most important thing people can do to avoid contracting diseases from wildlife, including bats, is to avoid handling them. Despite living in close proximity to 1.5 million bats in downtown Austin, Texas, no one has been harmed in more than twenty-five years.

Bat Rehabilitation

I (Barbara Schmidt-French) have cared for hundreds of bats over a period of many years as a licensed wildlife rehabilitator, and much of what I know about bats today is a result of this work. Caring for injured and orphaned bats in captivity is truly rewarding work, but it is time intensive. Moreover, treating microbats can be very challenging due to their tiny size. The work must be conducted under state regulations in

(continued)

Bat Rehabilitation, *continued*

cooperation with veterinary guidance and includes receiving pre-exposure rabies vaccinations.

I was trained and inspired by Amanda Lollar at Bat World Sanctuary, a nonprofit, all-volunteer organization recognized as a world leader in bat rehabilitation. Headquartered in Mineral Wells, Texas, Bat World has about twenty rescue centers throughout the United States. Each year Bat World rescues thousands of bats that might otherwise die. Lifetime sanctuary is given to non-releasable bats, including those that are orphaned, injured, confiscated from the illegal pet trade, or retired from zoos and research facilities. Veterinarians, biologists, scientists, and wildlife rehabilitators come from around the world to attend Bat World's annual rehab workshops, where I originally learned to treat bats with fractured bones, torn wings, lacerations, abscesses, eye infections, and parasites. I cleaned bats' teeth, clipped their claws, brushed their teeth, and, in an emergency, delivered bat pups by caesarian section. I raised orphaned pups on milk formula and watched them grow and become proficient at catching flying insects in flight cages.

As a result of this work, I have observed the behavior of some species of bats for thousands of hours in captivity and, in the process, contributed interesting information to the scientific community by collaborating with experienced bat researchers. But the greatest pleasure I get from this work comes when I am able to release a healthy bat back to the wild where it

Figure 23. An orphaned straw-colored flying fox (*Eidolon helvum*) sucks contentedly on a baby pacifier. *(Photograph courtesy of Bat World Sanctuary, www.batworld.org.)*

Bat Rehabilitation

belongs. Although we don't know what percentage of the bats we release survive in the wild, our work with John O. Whitaker Jr. of the Indiana State University Bat Center has proven that some hand-raised orphans can learn to feed on flying insects in large flight cages in captivity. It is unlikely that these orphans would have survived without the intervention of a rehabilitator.

Question 5: What is White Nose Syndrome?

Answer: Currently, the most pressing concern regarding bats is the death of thousands of insect-eating bats in the northeastern United States. During the winter of 2006–2007, bats near Albany, New York were first observed arousing from hibernation too early in the year, when it was still cold and there were no insects available for them to eat. They were found crawling in the snow and flying during the day, searching fruitlessly for food and perhaps for warmth. Large numbers of bats were found dead or dying in caves where they hibernate. Most of these bats had no fat reserves and many were found with severe damage to their wings, and scientists struggled with several questions. Why were the bats arousing too early? What happened to their fat reserves? Did they have enough fat stored up when they went into hibernation? Affected bats often had a white fungus growing on the wings and around the nose, hence the name that was given to this mysterious condition, White Nose Syndrome (WNS).

At least eight thousand bats died during the winter of 2006–2007. In June 2008, there was a three-day brainstorming session in Albany, New York, that focused on the mass die-off of bats in the Northeastern United States. According to a Bat Conservation International posting, " . . . thirteen scientists, including specialists in bat ecology and physiology, pathology, infectious diseases, toxicology, and environmental contaminants, presented their research findings and offered hypotheses for immediate study. More than eighty participants from two Canadian and twenty United States state and federal agencies, eight universities and four non-government organizations then discussed the existing knowledge and pending questions about the syndrome." Separate groups of scientists and wildlife managers met to identify the issues that were of most concern to them, and then everyone came together to define and prioritize the most promising research directions. The three questions that received top priority were: Why are affected bats starving? Are pathogens a direct cause of mortality? And are contaminants threatening either the bats or their food supply? Surveys were planned for caves and mines in the area to explore known WNS-affected sites and to identify possible new sites.

Since the original observations of bat deaths associated with WNS in New York, the syndrome has spread to bats in Massachusetts, Connecticut, Vermont, New Hampshire, Pennsylvania, New Jersey, and most recently as far south as Virginia. All of the hibernating species of bats in the region have been affected, including the little brown bat (*Myotis lucifugus*), the endangered Indiana bat (*Myotis sodalist*), the northern myotis (*Myotis septentrionalis*), the eastern small-footed myotis (*Myotis leibii*), the big brown bat (*Eptesicus fuscus*), and the tricolored bat (*Perimyotis subflavus*). Many of the bats die in hibernacula, while many of those that survive the winter later die during the months that follow due to the lack of fat reserves, serious damage to their wings, or weakened immune systems.

Perhaps as many as a million bats have died in just the past three years, and concern now focuses on the possible spread of the disease across the entire United States. According to Thomas Kunz at Boston University, the numbers and biomass of insects that could have been eaten in a single year by those missing million bats is staggering. He estimates that they could have consumed up to 1,388,912 pounds of insects! And although we are aware of many of the insect pests eaten by bats, new work by Elizabeth Clare, Erin Fraser, Heather Barid, and Brock Fenton of the University of Western Ontario in Canada demonstrates just how much more there is to learn. Using PCR (Polymorase Chain Reaction) and a sequence-based technique, the researchers were able to identify 127 different species of insects in the droppings of a single bat species, the eastern red bat (*Lasiurus borealis*). Although the eastern red bat is not a colonial species that uses caves or mines to hibernate as do bats currently affected by WNS, the new technique being used by Elizabeth Clare and her associates is likely to be used by others to identify more insect species in the diet of bats that are affected by WNS. These findings emphasize how the continued loss of huge numbers of bats across the United States could have a frightening impact on the delicate balance of nature.

Unless the cause of these deaths can be determined and scientists find a way to intervene, this is potentially the most devastating crisis to affect bats in the United States. Although scientists have no reason to believe that whatever is affecting these bats is something that can be transmitted to humans, they are concerned that people who visit caves may unknowingly transport the fungus from one cave to another, endangering bats in as yet unaffected hibernacula. Caves in the Northeast were closed in the winter of 2008–2009 to try to protect the bats from contracting the disease, and then they were opened in mid-May when the bats were no longer hibernating in the caves. In April 2009, a one year closing of almost all caves and mines on National Forest Service lands was imposed in the Eastern Region of the United States, and the United States Forest Service also closed caves and mines in the Southeast.

Numerous groups are now involved in the goal of finding a solution to WNS, including the United States Fish and Wildlife Service, the United States Forest Service, the National Park Service, the United States Geological Survey and National Wildlife Health Center, the Department of Defense, the Nature Conservancy, the Bureau of Land Management, the Tennessee Valley Authority and state wildlife divisions throughout the affected regions, Boston University, Bucknell University, Cornell University, Indiana State University, St. Paul's School, the University of Wisconsin, Indiana State University, Bat Conservation

International, the National Speleological Society, the National Wildlife Foundation, Defenders of Wildlife, the Bat World Sanctuary, the Northeast Bat Working Group, the Southeastern Bat Diversity Network, the Midwest Bat Working Group, the Western Bat Working Group, and others.

Scientists have now identified the fungus growing on affected bats as *Geomyces,* a fungus capable of surviving cold temperatures. They also find the fungus growing inside hibernacula where affected bats are found, and they do not yet know if the bats die as a direct result of the fungal infection, or perhaps as a result of the interaction between the fungus and some other aspect of the environment. Bat biologists, conservationists, virologists, bacteriologists, mycologists, pathologists, toxicologists, and physiologists across the United States are working diligently to determine the exact cause of these bat deaths.

Question 6: Do bats have enemies?

Answer: Because bats are capable of flight, they are able to easily escape many potential predators. However, bats do have numerous chance enemies, and individual predators sometimes learn to capture bats at the entrances to bat roosts. For example, the Puerto Rican boa constrictor, a large snake three to six and a half feet (one to two meters) long, hangs in a double-S pose from rocks or vegetation, poised to strike at the entrances of caves where bats live. At dusk, as clouds of bats swarm out into the night, the snakes lunge and try to capture bats in their mouth. If the maneuver is successful, the snake quickly coils around the bat's body and then inserts the bat down into its coils and swallows the bat whole while protecting its prize from other hungry snakes. Similar behavior has been observed among Cuban boas and in the United States among coach whip snakes such as those hanging at the opening of Bracken Cave in central Texas. Some snakes are able to climb cave walls to catch bats roosting on the ceiling, and other snakes have been observed preying on bats while the bats feed at flowers.

Birds of prey occasionally eat bats; falcons and hawks can sometimes be seen flying among swarms of bats when they emerge from their roosts at dusk, snatching bats out of the air. Owls have also been observed pursuing and capturing bats in flight. Roadrunners, blue jays, blackbirds, grackles, gulls, and crows have all been seen attacking bats during the day, particularly bats that roost in trees among the leaves. Louise Allen-Hristov (Boston University, personal communication) observed grackles pulling live bat pups out of crevices where they were roosting under a bridge in central Texas. The bat hawk of Africa and New Guinea feeds exclusively on bats, and even some carnivorous species of bats feed on small vertebrates, including other bats. Other animals that have been observed feeding on bats include opossums, skunks, raccoons, ring-tailed cats, bobcats, forest mice, stoats, civets, weasels, and centipedes, and there are even large spiders that have been observed eating bats that flew into their webs. Bat pups that are unable to fly for the first few weeks of life are particularly vulnerable to predators, and young bats that fall to the floor of a cave may be eaten by beetles, roaches, assassin bugs, and ants.

Even if a bat escapes from a predator, it may sustain broken or dislocated bones, torn wing or tail membranes, or internal injuries. Injured bats can die from infections or from injuries that prevent them from flying well enough to catch sufficient food. Injured bats that can't fly normally often end up on the ground where they are more vulnerable to other predators, including cats and dogs. Even healthy bats can be caught by domestic cats when the bats drop to feed on low-flying or crawling insects. Bat organizations have launched a public service campaign in the United Kingdom, urging pet owners in suburban and rural areas to keep their cats inside at night.

Question 7: How do bats avoid predators?

Answer: Bats are safe from most predators when they fly at night, and they are able to escape predators in the roost by simply flying away when they approach. But some bats hibernate, in which case they need a particularly secure location because they will not be alert for a sustained period of time. Similarly, many bats enter a state of inactivity (torpor) for a part of each day to conserve energy and are more vulnerable to predators at that time because they are not alert.

The main defense that bats have is choosing a safe roost. Sometimes their day roost is different from their night roost, but any roost, in addition to providing protection, must be relatively unobstructed. Roosts conceal the bats, protect them from being disturbed, and limit access by predators, whether the roosts are in buildings, caves, rocky crevices, or tree cavities. A suitable tree cavity needs to be high in the tree, without nearby branches that might facilitate access by predators or obstruct the bats' flight when they leave at dusk.

Some small bats roost inside furled leaves, and others roost under a tent they make of foliage. Bats that roost this way are alert when they are in the roost and have a good line of site from inside or under the leaves so that they can spot an approaching predator and fly away (see color plate B, Honduran white bats). Another defense is to create a group of tents; as a predator begins to approach one of the tents, bats in the other tents may be alerted and have time to escape.

Bat houses mounted in an appropriate location (see chapter 7, question 8: How can I build a bat house?) are another good solution for protecting bats from predators because they offer safe options in locations where good roosts may be scarce. If there are climbing snakes in the area, predator guards can be used to prevent the snakes from reaching the bat houses. A predator guard of this type is a two- or three-foot metal cylinder mounted around the post on which the bat house is attached. It is placed above ground level but below the bat house and has a screen closing off its top end, so a snake crawling up the post goes into the cylinder and is blocked from going further upward.

SEVEN

Bats and People

Question 1: Why are people afraid of bats?

Answer: There is no reason to be afraid of bats. However, a quick glimpse of a bat darting out from a tree at twilight could possibly startle a person—it is a natural response, especially to an animal that is rarely seen clearly. If we saw a cat under similar circumstances, we might also be startled momentarily, but because cats are familiar and we know what to expect from them, most people would not be frightened. Bats are relatively unfamiliar, and they are thought of by many people as strange or even menacing creatures. The image of the bat's dark wings draped around its body when it is at rest led to a common belief that witches transformed themselves into bats, so bats were sometimes nailed to barns to repel witches. In French, the word for bat is *chauve-souris,* literally, "bald mouse," an association with an animal that many people are not pleased to see nearby. In fact, the shy and reclusive bat is more closely related to people than it is to rodents.

You've heard the expression "like a bat out of hell"? Caves have always been seen as mysterious links to the bowels of the earth, and the image of a mass of bats streaking out of the mouth of a cave into the darkness, forming long ribbons that snake across the night sky, is quite a unique sight and, understandably, can inspire fantasy. It can take an hour or more for the bats to fly out of a cave if it is a very large colony. People gather nightly near bridges or caves where large numbers of bats roost, mesmerized as they watch the strange spectacle of bats emerging at dusk.

Bats commonly roost in places considered mysterious or even haunted, such as tombs, old abandoned mines, vacant buildings, and church steeples. You've heard of "bats in the belfry" or "going batty," equating bats somehow with insanity? Some cultures take this a step further into the realm of the supernatural and refer to bats as creatures created by the devil, representing danger and death. In the early 1600s, William Shakespeare used bat's fur, along with parts of other animals that in his day symbolized evil, as part of the witches brew in *Macbeth:*

Eye of newt, and toe of frog,
Wool of bat, and tongue of dog,
Adder's fork, and blind-worm's sting,
Lizard's leg, and howlet's wing,—
For a charm of powerful trouble,
Like a hell-broth boil and bubble.
(*Macbeth,* IV, 1, 14–15)

Bats have also been described as human souls that leave the body at night and return in the morning, or as the transformed bodies or souls of dead people who are not at peace, prowling the night and, in some cases, sucking the blood of human victims. Hollywood has certainly contributed to this frightening image of bats with a slew of films and horror stories that have exploited the association of these gentle and shy creatures of the night with danger and death. Gary McCracken (University of Tennessee at Knoxville) has traced the association of bats with vampires to stories that were reported in Europe in the sixteenth century by explorers, one of whom described his toes being bitten by bats when he was asleep in the area that is now Costa Rica. He was probably bitten by a vampire bat, found in the New World tropics, which was totally strange and unheard of in Europe. Bram Stoker gets credit for connecting the foreign image of vampire bats with the eastern European vampire stories in his 1897 novel, *Dracula.* In reality, there are only three species

Folklore: A Selection of Bat Legends

Gary McCracken of the University of Tennessee at Knoxville has collected all sorts of folkloric information about bats. The legends and folktales that he and others have gathered from different cultures provide interesting insights into the way people think about bats and how they try to make sense of animal behavior that seems unusual or abnormal to them. For example, since many creatures are active during the day and sleep at night, the idea that bats are creatures of the night holds a certain mystery that begs to be explained. The ancient Hebrew storyteller Aesop tells a tale that focuses on shame as the reason why bats are nocturnal. In his fable, a bat borrows money for a business venture that fails, so the bat hides during the day to avoid its creditors. We've extracted a few other examples from McCracken's writing to give you a taste of some other attempted explanations.

There is an Islamic legend claiming that a bat was created by Jesus. During the fast of Ramadan, eating is forbidden between sunrise and sunset, but Jesus could not see the setting sun from where he lived because his view was obscured by mountains. He created a bat out of clay and breathed life into it, and it flew away into a nearby cave. Every evening at sunset it emerged from the cave, letting Jesus know that it was time to break the fast.

Another legend contains a different explanation of why bats fly at night. The Kanarese of southern India tell a story about bats once being unhappy birds. They went to the temples and prayed to be turned into humans, and their prayers were partially answered; they were given hair, teeth, and human faces, but otherwise remained like birds. After that, they were ashamed to meet other birds because they looked so strange, so they fly at night to avoid being seen by day-flying birds. The bats return to the temples each morning to pray to be turned back into birds.

Some South American legends portray the bat as a heroic

(continued)

Folklore: A Selection of Bat Legends, *continued*

figure. A folktale from northern Argentina describes the leader of the very first people as a hero bat or bat-man who taught people everything they needed to know. A Brazilian myth has a similar slant, describing a bat-hero leading the people through the night, guiding them toward the light. And, of course, in the United States we have our own Batman, a reclusive and secretive superhero who lives in a bat cave and drives a bat mobile.

American Indians have various legends that focus on how bats were created from other animals. In a Cherokee fable, two small mouse-like creatures wanted to participate in a ball game in which the birds challenged the animals. Because they had four feet, the small creatures asked if they could play with the other animals. The larger animals made fun of how small the creatures were and chased them away. Then they asked the eagle, the captain of the bird team, if they could play on the side of the birds. The birds took pity on them and fashioned wings for them out of the skin of a drum. They stretched the leftover pieces of skin between the forelimbs and the hind limbs of the creatures, making the first flying squirrels. These new flying creatures helped the birds win the ball game.

of true vampire bats and they are found only in the New World tropics (see chapter 1, question 14: Do bats drink blood?).

How about "blind as a bat"? The misconception that bats are blind causes some people to fear that bats are likely to fly into them, although in reality bats see quite well. The belief that bats will get tangled in a person's hair is an unfounded fear, perhaps due to the misconception that bats can't see. It may also be due in part to the feeding style of some bats that swoop down to prey on insects drawn to lights where people gather, which may give the impression the bats are drawn to the people. Stephen

Kellert and Edward Wilson assert the fear of bats becoming entangled in a woman's hair is rooted in the New Testament. Paul proclaimed that women were required to cover their heads in church to protect them from dangerous spirits with demonic powers that were attracted to unbound female hair. This developed into the belief that bats were irresistibly attracted to women's hair and, according to Richard Cavendish, once entangled there, could only be released by a pair of scissors wielded by a man.

These enigmatic flying mammals are, indeed, unusual creatures, but they are certainly not the vicious, blood-thirsty demons portrayed in the movies. Fortunately, in many cultures, bats have a more positive image (see folklore sidebar); in fact, where diurnal bats are common, they are considered lucky omens, probably because they are familiar and do not have the associations with night and death. Public education about the value of bats to the economy is helping to improve their image (see this chapter, question 3: Are bats useful to plants? and question 13: What is being done to protect bats and how can I help?).

Question 2: What dangers do bats face from people?

Answer: Bats are threatened by human activities in a number of ways. Despite protective regulations, people still sometimes kill bats in roosts due to unfounded fears, and entire bat colonies have been destroyed by humans who have altered cave systems either intentionally or in error. Bats require very specific temperatures and humidity levels, especially when they are raising young and hibernating, so minor changes in the physical structure of a cave roost can result in temperature and air-flow changes that render it unsuitable for the bats that use it. Also, hikers or others who enter caves containing hibernating bats may awaken them, causing them to use up valuable fat reserves that they need in order to last through the winter (see chapter 3, question 9: Do bats hibernate?). Disturbing a maternity colony can cause mothers to panic and abandon their young. Bats that roost in abandoned mines sometimes become trapped when the mine is closed off or filled in with dirt to prevent people from entering. Logging or cutting of snags (dead trees that are still standing) can create a housing shortage for bat species that roost in trees. Due to a loss of habitat, some bats now roost in buildings, and in many places bats roosting in buildings are simply exterminated. Unwanted bats in buildings should be evicted, not exterminated (see this chapter, question 7: How can bats be safely evicted from a building?).

Thousands of migratory bats are killed each year by wind turbines, those slender propellers mounted on tall poles, which are becoming an increasingly common alternative source of energy production. At several sites in North America large numbers of dead bats, including the highly migratory hoary bats (*Lasiurus cinereus*), eastern red bats (*Lasiurus borealis*), and silver-haired bats (*Lasionycteris noctivagans*), have been found at the base of the turbines. For example, an estimated two thousand bats were killed during just one six-week period in 2004 at sixty-four turbines studied in Pennsylvania and West Virginia. An interesting study was reported in 2008 by Erin Baerwald (University of Calgary in Alberta, Canada) and colleagues. The study found that death from "barotrauma" is likely one of the causes of bat mortality at wind turbines. The rapid drop in air pressure near the moving blades of the turbines results in an over-expansion of air sacs in the lungs, which causes surrounding capillaries to burst. Only half of the bats they examined showed evidence of injury from the turbine blades, but 90 percent showed signs of barotrauma. The Bats and Wind Energy Cooperative (BWEC), led by Bat Conservation International (BCI), is currently collecting data at wind turbine sites to better evaluate risks and find solutions to prevent bat kills.

Thanks to conservation programs, some mining companies are now working to protect bat roosts in abandoned mines by covering their entrances with gates, or by surrounding the entrances with fences that keep people out but allow the bats to come and go freely. For example, Bat Conservation International is leading the Death Valley Mine Closure Alliance in collaboration with the National Park Service, Rio Tinto Minerals, and the California Department of Conservation. Together, these organizations are protecting valuable bat habitat by gating abandoned mines on Rio Tinto and National Park Service property in Death Valley National Park. The Indiana State University (ISU) Center for North American Bat Research and Conservation is working with the Indiana Department of Natural Resources to gate and study the results from gating of abandoned mines in that state. Good forest management practices are also being used in some areas to ensure that enough trees are left intact during harvesting to provide sufficient roosts for bats. And finally, more and more pest-control operators are learning to evict bats from buildings in a process called exclusion, which uses safe eviction techniques (see this chapter, question 7: How can bats be safely evicted from a building?). Together, all of these conservation efforts can make a difference for America's bats.

Question 3: Are bats useful to plants?

Answer: Most bats either eat insects or fruit or drink nectar from flowers, and some species consume more than one of these types of food. This makes bats exceptionally useful to plants, which provides significant benefits to the economy. Insect-eating bats are economically important because they help to control crop pests like corn earworms, armyworms, and cucumber beetles, to name just a few of the insects they eat in large quantities. Because they naturally diminish the populations of these pests, the number of pesticide applications required to keep crops healthy is reduced.

In 2008, Margareta Kalka and Adam Smith (Smithsonian Tropical Institute) and Elizabeth Kalko (University of Ulm in Germany) reported results from their research in Panama: they enclosed plants only during the day to exclude birds and only at night to exclude bats, and their results demonstrated that bats play an important role in protecting plants from insects. Plant-eating insects account for more than 70 percent of the diet of the little big-eared bat (*Micronycteris microtis*) that was present in the area of Panama where they did their research. Over the ten-week study period, they found that plants that bats could not reach had 153 percent more insects per 10.8 square feet (per square meter) than uncovered controls, and plants that birds could not reach had 65 percent more insects than uncovered controls.

They also did a six-week study at an organic cacao farm in Panama where they regularly placed caterpillars on platforms of cacao leaves attached by wires to branches and then covered the area during the day or during the night. When the caterpillars were exposed to bats at night, predation rates were more than three times higher than when they were exposed to birds during the day. A related experiment using plant enclosures was conducted in Chipas in southern Mexico by Kimberly Williams-Guillen, Ivette Perfecto, and John Vandermeer from the University of Michigan at Ann Arbor. In 2008, they reported that their study also showed that bats played a significant role in reducing crop pests, although in that area the presence or absence of migratory birds caused more variation in the results.

Nectar bats usually drink from night-blooming flowers, many of which are bell-shaped or shaving-brush shaped or have wide corollas that provide good guides for bat snouts so that they can easily reach the nectar and come in contact with the plant's pollen. Many of these flowers are especially attractive to bats because they have a unique scent or are acoustically conspicuous, which means they easily reflect echolocation calls used by some bats to locate food. As the bat drinks nectar from a flower, pollen grains (male gametes or chromosome-carrying cells) adhere to the bat's head. When the bat visits one flower after another, the pollen grains are transferred to the female ovum (gamete) of the next flower, leading to fertilization, with the result that the plant reproduces by bearing seeds or fruit. Without pollination, the plants would be sterile and there would be no fruit.

From the southwestern United States to Central America, the lesser long-nosed bat (*Leptonycteris yerbabuenae*) feeds on nectar and pollen from a number of cactus plants (see color plate D). The two other species of nectar bats in the United States are the

Mexican long-nosed bat (*Leptonycteris nivalis*) and the Mexican long-tongued bat (*Choeronycteris mexicana*). These nectar-feeding bats also include some fruit in their diet. In addition, another fruit-eating bat, the Jamaican fruit bat (*Artibeus jamaicensis*), is occasionally seen in the United States in the Florida Keys, where it may feed on sea grapes, sea almonds, or exotic fruits. Bats in the families Phyllostomidae (New World leaf-nosed microbats) and Pteropodidae (Old World megabats) pollinate many plants in tropical parts of the world (see sidebar on pollination).

There are many tropical species of fruit-eating bats (also called flying foxes) in Africa, Asia, Australia, and the islands of the Indo-Pacific ocean. Some of the large flying foxes eat fruits with very big seeds, like mangos, and they often take the fruits to a place distant from the tree where it grew. In the process of eating the fruit, they drop the seed to the ground where it can start a new tree.

Fruit-eating bats are big eaters, and when dwarf epauletted fruit bats (*Micropteropus pusillus*) were provided with unlimited amounts of fruit in flight cages as part of Donald Thomas's research in West Africa, they ate as much as two and a half times their body weight in a single night. Buettikofer's bats (*Epomops buettikoferi*) ate up to one and a half times their own weight under the same conditions. Some bats digest their meals in as little as fifteen minutes, and if they've eaten fruits with small seeds, the seeds pass rapidly through their digestive system and out in their droppings as they fly.

Thomas put sheets of plastic on areas of cleared land in Cote D'Ivoire, West Africa, to collect the seeds dropped during the night by flying fruit bats. He calculated that each square meter received anywhere from one to hundreds of seeds each year, almost all of them excreted by bats. Seeds that were spit out were partially crushed and could be differentiated from the ones that had passed through a bat's digestive system. Very few seeds were collected during the day, and those were presumably dropped by birds. Birds generally stay in a tree longer when they're feeding and drop most of their seeds right next to the tree, where insects and other seed predators tend to gather and eat the seeds.

Thomas collected some of the seeds dropped during the night by bats and took them back to his lab to see how quickly they would germinate compared to seeds removed directly from ripe fruit. He found that the seeds that had passed through a bat germinated first; after six days, more than half of them had sprouted, and within a few more days they had all germinated. He observed the seeds for three months and found that no more than 10 percent of the seeds he had extracted from ripe fruits ever germinated, and few of the partially chewed seeds ever sprouted. Clearly, the enzymes in a bat's digestive system prepare the seeds to germinate.

Detlev Kelm (University of Erlangen in Germany) did related research with K. Wiesner, Otto van Helversen, and others, installing artificial roosts to attract bats to deforested areas in Costa Rica where only fragments of vegetation remained. They collected "nocturnal seed rain," seeds dropped from flying bats on fine nylon mesh stretched over frames, and found that seed dispersal and new plant growth (revegetation) were significant around the roosts. This suggests that attracting bats can provide an exciting alternative to extensive (and expensive) replanting for restoring habitat that has been deforested.

Fruit and nectar bats either pollinate or disperse the seeds of an estimated five hundred species of tropical trees and plants around the world. Propagation of plants by bats is vital to the regeneration of rain forests in areas that have been cleared due to slash-and-burn agriculture or indiscriminate logging, and the range of products that come from plants that survive because of their interaction with bats is astounding. Flying foxes pollinate the coconut palm, used in the production of coconut oil, margarine, soup, flavorings, soap, detergents, bath gels, the foaming agent in toothpaste, components used in perfumes and cosmetics, as well as lubricating fluids for racing cars and airplane engines. And that is just one species of plant! Some flying foxes pollinate certain species of mangrove trees, and mangrove wood is used for charcoal, wood chips, dock pilings, scaffolding, and fishing boats; other parts of trees are used in the production of tannin, textile fibers, cosmetics, incense,

Pollination

In order for most plants to grow fruit, there must be a way for them to cross-fertilize, which means that male chromosome-carrying cells (gametes) must somehow travel from a plant to fuse with the female gametes in another plant—this process is called pollination. About 80 percent of all plants are pollinated by living organisms (biotic pollination). Pollinators include bees, butterflies and other insects (entomophily), and vertebrates such as birds and bats (zoophily). The pollen of a few plants is transported in water and, for the remaining species, by the wind.

Gametes are delicate single cells, and male gametes are enclosed in pollen grains. An animal is attracted to a flower by its fragrance, color, and shape, all of which promise a reward of nectar or pollen. A few plants mislead animals into visiting with the promise of sex by using chemical bouquets that mimic the inviting scent of a receptive female. For most animals, collecting pollen is not their goal, but in the process of obtaining nectar, the flower's pollen grains adhere to the animal's head or body. When the animal visits the next flower, some pollen grains rub off onto the flower's stigma, the receptive part of the plant containing the female gametes (ovule). This results in fertilization, which makes it possible for the plant to reproduce by bearing fruit.

Pollination by bats (chiropterophily) is generally limited to the tropics and subtropics and is almost exclusively done by two families of bats, Pteropodidae (Old World megabats) and Phyllostomidae (New World leaf-nosed microbats). These species tend to have broad wings that allow them to hover while they drink nectar, although some species do land and perch while feeding. A nectar-drinking bat typically has a long nose and a long tongue to reach the fluid deep inside certain flowers. They may also have tiny hairs on the end of the tongue that are used to lap up the nectar. The exceptional tube-lipped nectar bat (*Anoura fistulata*) has a tongue that can

(*continued*)

extend twice the length of its body. This bat has one species of flower, *Centropogon nigricans,* all to itself as a nectar source because the flower's corolla tubes are so long that no other pollinator can reach its inner juice. Mangos, bananas, guavas, plantains, avocado, dates, and breadfruit are just a few of the hundreds of plant species that are dependent on bats for pollination.

Many plants have evolved flowers that are well-suited to being pollinated by bats. These plants have night-blooming flowers that are often large and firm with wide corollas, usually white or pale in color, although some are more colorful. Some flowers reflect the calls of echolocating bats, making them acoustically conspicuous and giving them an advantage in attracting bat pollinators. Approximately two-thirds of Mexico's seventy species of columnar cactus produce flowers that attract bats as well as bees, hummingbirds, and nocturnal hawk moths—large moths that are sometimes mistaken for hummingbirds. Long-nosed bats (*Leptonycteris* sp.) in the southwestern United States pollinate the bell-shaped flowers of agave (*Agave augustifolia*), saguaro (*Carnegiea gigantea*), cardon (*Pachycereus pringlei*), and organ pipe (*Stenocereus thurberi*) cacti. The cacti's flowers appear before their protective spines develop, leaving the plants accessible for a period of time during which they can be safely pollinated.

Bats are attracted to flowers with shapes that provide a good guide for bat noses. Some flowers are shaped like a shaving brush, like those of the African baobab (*Adansonia digitata*) and two other baobab species (Brevitubae) native to Madagascar. (The remaining five species of baobabs in Madagascar and Australia are pollinated by long-tongued hawk moths.) Other plants with similar brush-type flowers are myrtles (Myrtaceae), plants bearing sapotes or fruits (Sapotaceae), and mimosoid legumes (shrubs and trees) (Mimosaceae). There are also bat-pollinated plants that have clusters of flowers with showy stamens but no petals, and some that have eas-

Pollination

ily reached flowers at the top of the canopy or growing on the trunk or lower branches. Bats have easy access to pendulous clusters of flowers that hang below most of the foliage, like the flowers of the sausage tree (*Kigelia pinnata*), endemic to Africa.

musical instruments, dyes, thatch, matting, sugar, alcohol, cooking oil, vinegar, honey, paper, and medicines.

Question 4: Do people eat bats?

Answer: On some South Pacific islands and in other parts of the Old World, bats are considered a delicacy as well as an aphrodisiac. They are also used as an ingredient in remedies for blindness, baldness, and a host of other problems. At one time, hog raisers on the islands caught and cooked the local bats to feed to their livestock. By 1974, only one hundred Mariannus flying foxes (*Pteropus mariannus*) remained on the island of Guam, and by 1978, another local species (*Pteropus tokudae*) had been hunted to extinction. Huge numbers of bats, as many of twenty-nine thousand per year, were imported to Guam for human consumption, and Merlin Tuttle, the founder of Bat Conservation International, reported that in 1983 a bat dinner on Guam might have sold for as much as twenty-five dollars a plate.

After World War II, an unusually high incidence of two neurodegenerative disorders, Amyotrophic Lateral Sclerosis and Parkinsonism-Dementia Complex, (ALS and PDC), sometimes coexisting in the same person, was reported among the Chamorros, the local people living in Guam. The fact that the diseases were rare among Chamorro populations on other islands suggested that environmental factors were the cause, and the consumption of fruit bats, a traditional food on the island, was thought to be the cause of the illness. The theory was that the bats ate some fruits with seeds that contained toxins, and the toxins would become concentrated in the bat's body and then be passed on to the people who ate the bats, causing the diseases. In 2007, research by Amy Borenstein (University of South Florida at Tampa) was not able to definitively establish an association between eating fruit bats and the diseases.

In 2002, Chris Plato (University of California at La Jolla) and his team of researchers conducted a forty-year follow-up study of the Chamorros people. They found that relatives of patients with the illnesses are at a significantly higher risk for developing the diseases than Guamanians whose relatives are disease-free, but they found no link to eating bats. Molecular analyses done by Meredith Hermosura in 2005 (University of Hawaii at Manoa) and by another group of investigators in 2004 led by Huw Morris (Manchester Metropolitan University in the United Kingdom) did not find a clear genetic link that would establish cause and effect, but they did find that people with a particular genetic makeup are three times as likely to succumb to the disease as people without that profile. But again, they did not find a link between susceptibility to the disease and a history of eating bats. Meanwhile, the bats have been classified as endangered due to overharvesting, and the incidence of the diseases has declined over the past fifty years. There is obviously more work to be done to understand the genetic and environmental causes of these illnesses.

Question 5: Do bats have any scientific or medical value?

Answer: Scientists study a wide range of plants and animals to determine if their natural properties can be beneficial to humans, and some researchers have turned their attention to bats. The clot-dissolving enzyme in the saliva of the common vampire bat (*Desmodus rotundus*), which allows blood to keep flowing while the bat is feeding, is 150 times more powerful than the thrombolytics, or anti-clotting drugs, that are currently on the market. A synthetic copy of the enzyme, DSPA (*Desmodus rotundus* salivary plasminogen activator), was recently developed and is now being tested in clinical trials in the hope that it will lead to the development of safe new medications that will save lives and limit the damage caused by blood clots.

Studies of echolocation in bats have contributed to the development of navigational aids for the blind. They have also led to the development of software that emulates bat echolocation for use by the U.S. Navy to detect and distinguish between submarines and explosive mines. And engineers at the University of Arizona received a grant from the U.S. Air Force in 2008 to try to develop "cognitive radar," a complex system that can change its waveforms in response to environmental cues, much like the sophisticated "cognitive sonar" used by echolocating bats (see chapter 4, question 1: How does echolocation work?)

Another interesting study has implications for aerodynamics and surveillance specialists. A team of scientists from Sweden and the United States recently studied Pallas long-tongued bats (*Glossophaga soricina*) as they fed on nectar in a wind tunnel. Using fog, lasers, and high-speed cameras, they observed that as a bat flapped its wings downwards, the motion of the wings created a tiny cyclone of air known as a leading edge vortex, which appears to provide as much as 40 percent of the lift the bat needs to hover while it laps nectar from a flower. Applications of this discovery may be used to improve the design of miniature surveillance airplanes.

Bats are important to the economy as pollinators and seed dispersers and as a natural system of insect control (see this chapter, question 3: Are bats useful to plants?). Even waste products of bats are valuable as a natural fertilizer. In several countries, bat guano (bat feces) is regularly collected from caves and used by local farmers (see chapter 3, question 7: What is bat guano?).

Question 6: Are any bats endangered?

Answer: Edward O. Wilson, the distinguished evolutionary biologist, is often quoted blaming humans for "nearly every loss of species in the past few thousand years." Some paleontologists believe that we are on the verge of the seventh great extinction, and while prior extinctions were likely due to cosmic accidents or climate change, the present ecological crisis is generally attributed to the destructive impact of human activity on the earth's terrestrial and aquatic ecosystems. For example, although bats provide valuable services to the environment by consuming large volumes of insect pests and by pollinating plants and dispersing seeds, in some parts of the world bats are routinely shot or smoked out of caves simply because they are disliked or feared. In Latin America, bats are often killed in cave roosts due to the false belief that all bats are vampires.

Habitat destruction and intentional or unintentional killing of bats threatens many species worldwide. More than twenty years ago, Merlin Tuttle reported that tens of thousands of megabats called African straw-colored flying foxes (*Eidolon helvum*) were being slaughtered and imported to Guam and Saipan for food. They are considered a delicacy in those areas, and people at that time were paying from ten to twenty-five dollars or more per meal. Local farmers also killed the bats because they were considered pests in the region's mango plantations. And because bats, in general, were disliked by the growers, other species were killed as well; so whether as pests, delicacies, or innocent relatives, bats have been methodically slaughtered, and several species of flying foxes have become extinct as a result of overhunting and inter-island trafficking.

Some species of bats are considered "keystone" species—that is, species whose behavior is central to their ecosystem. If, for example, the Egyptian fruit bat (*Rousettus aegyptiacus*) and other bats that pollinate the baobab tree were to be eliminated, "the loss could trigger a cascade of linked extinctions," according to Norman Myers. About one-third of all bat species in the world have not been evaluated as to their conservation status, and only about 40 percent of those that have been evaluated are classified as having stable populations. The remaining species have listings ranging from "potentially vulnerable" to "probably extinct." Twenty-nine bat species were included in the most recent International Union for Conservation of Nature (IUCN) register of critically endangered bats (2003), as listed on the Web site http://www.earthlife.net/mammals/bat-critical.html. Eight species of bats in the United States and its territories are currently listed as endangered by the U.S. Fish and Wildlife Service, including the species listed below.

Gray bat (*Myotis grisescens*)
Indiana bat (*Myotis sodalis*)
Ozark big-eared bat (*Corynorhinus [=Plecotus] townsendii ingens*)
Virginia big-eared bat (*Corynorhinus [=Plecotus] townsendii virginianus*)
Lesser long-nosed bat (*Leptonycteris yerbabuenae*)
Mexican long-nosed bat (*Leptonycteris nivalis*)
Hawaiian hoary bat (*Lasiurus cinereus semotus*) (Hawaii)
Guam flying fox (or little Mariannus flying fox) (*Pteropus tokudae*) (Guam)
Mariannus flying fox (*Pteropus mariannus mariannus*) (Guam)

Fortunately, organizations like Bat Conservation International continue working to protect important bat habitats through direct conservation action, by supporting research, and by educating people about the essential roles bats play in the environment (see this chapter, question 13: What is being done to protect bats and how can I help?).

Question 7: How can bats be safely evicted from a building?

Answer: Because bats are losing wild habitat due to human activity, some species now roost in structures built by people, including buildings, bridges, and road culverts (large pipes that conduct water under walkways or roads). Bridges can provide good roosting places for bats. The Congress Avenue Bridge in Austin, Texas, has become a tourist attraction where people gather daily at dusk during the summer to watch bats streaming out from under the bridge on their nightly search for food.

Although bats do not have the kind of teeth that can chew through building materials like rodents do, bats can sometimes become a noise or odor nuisance when large colonies live in a home or other building. It is inhumane and, in some places, including the United States, illegal to use poisons to exterminate bats because poisons increase the likelihood that sick bats will end up on the ground where they could have contact with people or pets. Besides being inhumane and dangerous, this

Figure 24. A male Townsend's big-eared bat (*Corynorhinus townsendii*) roosts in an abandoned building. *(Photograph courtesy of Michael Durham, www .DurmPhoto.com.)*

practice does not prevent bats from re-inhabiting the roosts at a later date. To avoid this, bats should be evicted from buildings by using proper "exclusion" techniques as described below.

Bats can enter a building through torn window screens, chimneys, cracks or holes in siding or soffits, unscreened louvered vents, or under flashing that has separated from the building. Any place where building materials have shrunk, warped, or moved apart provides a possible entry hole for bats. Bats can be safely evicted from buildings by covering access holes with flexible netting or plastic tubes that act as one-way valves, allowing bats to get out, but not back into the building. Netting should have a mesh of one-sixth inch or less and should be secured over the opening on the top and sides, but left to hang open several inches below the opening. Bats will come out through the hole and crawl around until they find their way out from under the netting where it is left open on the bottom. When the bats return, they tend to fly straight to the screened opening and do not usually figure out that they need to go back to the bottom of the netting to reenter.

Tubes placed over openings used by bats should be smooth inside, have a diameter of about two inches, and be pointed downwards so bats can slide down and out through the tube. Because bats are not able to cling to smooth surfaces, they cannot climb up the tube to get back in. The netting or tubes placed over openings used by bats are left in place for about one week to make sure all bats have gotten out of the building, and then the openings are permanently sealed shut to prevent the bats from getting back in. This procedure is called exclusion.

Exclusion of common house-dwelling bats such as the little brown bats (*Myotis lucifugus*) and big brown bats (*Eptesicus fuscus*) should be done after young are flying or, in some cases, after the bats have left for the winter. Little brown bats migrate to caves to hibernate, although big brown bats often hibernate in small numbers in buildings. Do not exclude bats roosting in buildings during cold winter months. Bats should not be excluded during the time when they are giving birth and raising their babies because bat pups that are not yet able to fly would be trapped inside. Instead, bats should be excluded in early spring before young are born or, better still, in the late summer after the young are able to fly and feed on their own. Keeping buildings in good repair goes a long way toward preventing bats from taking up residence, and to provide an alternate roost, people sometimes put up bat houses before they evict the bats from a building (see this chapter, question 8: How can I build a bat house?).

Question 8: How can I build a bat house?

Answer: Bat houses, also called bat boxes, have been used in Europe since the early twentieth century, but they have only been used in the United States since the 1980s. Providing additional roosting sites for local or migrating bats can be a valuable contribution to your neighborhood, and if you provide a warm, safe place for bats, they may repay you by reducing your insect population. Bat Conservation International's Bat House Project has been quite successful, and with the exception of areas where there are extreme temperature variations between day and night, 85 percent of the bat houses that were properly constructed and mounted have been occupied by bats. In the United Kingdom, volunteers have been very active installing thousands of bat houses, although they report a lower occupancy rate.

So what are the secrets to a successful bat house? You can build the house yourself from detailed instructions available in several books, or you can order one ready-made from the list of approved vendors on the Bat Conservation International (BCI) Web site. A bat house looks somewhat like a bird house, except that it has three-quarter-inch to one-inch slats that are open at the bottom where the bats enter. If you build a house yourself, make sure the wood you use has not been treated with preservatives or chemicals that might irritate the bats. According to BCI, bat houses should be at least two feet tall, have chambers at least twenty inches tall and at least fourteen inches wide, and have a landing area extending below the entrance from three to six inches. All partitions and landing areas should be roughened (scratched or grooved horizontally) so that the bats can get traction, and the houses should be well sealed with caulk to eliminate drafts and leaks. Paint color also seems to be important; darker colors work better in cooler climates, while lighter colors work best where it is warmer. You can experiment by installing two bat houses, each painted a slightly different shade, to see which color the bats prefer. The house should be placed on a post or building twelve feet or higher above the ground to make sure the bats have a clear, unobstructed flight path into the bottom entrance. The bat house should be in a warm, sunny spot, preferably facing south or east, and within a quarter mile of water.

Question 9: How do scientists capture bats so they can study them?

Answer: In order to study bats in the laboratory, or to measure or mark them for observation in the field, researchers need to capture wild bats in nets or traps designed to catch them without causing injury. Biologists set up nets and traps in places where bats let down their guard or are otherwise vulnerable, for example, along a well-used path where bats may not always employ echolocation because it is familiar territory. One common method of capturing bats is use of mist nets, which are made of fine black nylon or polyester mesh suspended between two poles. Sometimes nets are set close together at angles, so even if the bats detect and avoid the first one, they may not detect the second one in time to avoid it. As bats fly into the nets, they become entangled in the fine mesh. The nets should be monitored continuously and bats removed quickly to prevent them from injuring themselves as they struggle to get free.

A harp trap, also known as the Tuttle trap (named after bat researcher Merlin Tuttle), consists of vertical threads of nylon fishing line or other smooth material, strung on a frame with a bag or container below. When bats fly into the lines, they slide down into the container, which is designed so that most bats cannot escape. This trap is best used at an entrance to a cave or mine.

Scientists must obtain permits from state or federal wildlife agencies that regulate the capture and use of bats for scientific purposes, and every effort is made to protect the bats from trauma and injury. Captures can be limited in various ways by regulations that restrict the number of bats that can be taken for a particular scientific study, or the reproductive status of the bats that can be captured may be limited. For example, some regulations prohibit the taking of females that are pregnant or lactating. Researchers must also follow the regulations set forth by animal care and use committees that oversee the captive care of wildlife used in scientific studies. Bats under these conditions provide scientists the opportunity to study bat behavior and other aspects of their life history.

Question 10: Can bats be domesticated?

Answer: Trained and licensed bat rehabilitators like co-author Barbara Schmidt-French keep bats in domesticated situations when they are caring for individuals that have been injured or orphaned, and some species quickly adapt to captive life under those conditions. Individual bats may even appear to enjoy interacting with their caretaker.

Some bats can be taught to exhibit specific behaviors on cue, for example, Rachel Page trained the fringed-lipped bat (*Trachops cirrhosus*) to fly to food in captivity in response to hearing recorded calls of frogs or toads. The bats quickly learned to associate food with specific calls, even when those calls represented toxic prey they would have avoided in the wild. But like other wild animals, bats do best in their natural habitats, and most species of bats do not fare well in captivity. There are a few exceptions, like Jamaican fruit bats (*Artibeus jamaicensis*), which appear to thrive and reproduce in zoos, yielding multiple generations of captive-born offspring. Although some fruit-eating bats may fare well in zoos, insect-eating bats are much more difficult to keep alive and healthy in captivity because it is very difficult to supply them with an adequate diet.

Figure 25. *Artibeus jamaicensis*, a Jamaican fruit bat, munches on a peach at the Bat World Sanctuary. *(Photograph courtesy of Bat World Sanctuary, www.batworld.org.)*

Many colonial bats require the companionship of other bats of their species in order to thrive, and forcing a colonial bat to live a solitary life in captivity would not be humane. Bats also require large flight areas and very specific temperature and humidity levels, and in most cases, the humidity levels they prefer are much higher than those that humans find comfortable. And bats, like other wild animals, are susceptible to rabies (see chapter 6, question 3: Do all bats have rabies?), which is another reason why they should not be sought out as pets. Bat care is extremely time intensive, and other than when a bat is in the care of a trained professional rehabilitator, educator, or researcher, it is not advisable to try to keep bats in captivity.

Question 11: How can I photograph bats?

Answer: As previously discussed, it is not advisable to try to catch or handle bats for any reason, including photography, unless you have been trained to do so and have had rabies vaccinations. Bats are difficult to photograph because most of them are very small, shy, fast-moving, and only active at night. Many of the common photographic images of bats show them snarling at the camera, but they probably look that way because they felt threatened by the activity of the photographer.

Researchers and professional photographers who photograph bats for study use elaborate equipment that is beyond the scope of the amateur. A tripod, fast film or a fast digital camera, a camera with an infrared trigger, a good flash-metering system, and as many extra flashes as one can afford are the minimal requirements. In some instances, Merlin Tuttle, known for his extraordinary bat photos, has used as many as eight additional flash units to properly illuminate a setting. He has also worked with caged bats that have become accustomed to his presence, even trained by him, and he shoots hundreds of frames to get one good shot. He has written that the typical equipment he takes with him when he goes into the field to photograph bats weights approximately 350 pounds. Scott Altenbach, from the University of New Mexico in Albuquerque, is another expert bat photographer and has written about the many aspects of the art and science of obtaining clear images of these fast-moving, nocturnal subjects.

Question 12: What is a bat detector?

Answer: Bat detectors convert inaudible, high-frequency bat calls (over 20 megahertz) into low-frequency sounds that humans can hear, allowing researchers and others to locate and listen to bats in the dark. A bat detector basically consists of a microphone to collect the sound, a circuit that converts the sound to a lower pitch, a speaker to broadcast the sound, a battery or other power source, and usually a recording device to preserve the calls for later study. Researchers study the acoustic structure of bat calls to learn many things, including how the frequencies used by a bat change over time.

There are different types of bat detectors. A modern heterodyne bat detector mixes the input signal from the bat with signals from two oscillators, and it can be tuned to a chosen frequency so that it will pick up the calls of bats that echolocate within a given range above and below that frequency. A frequency-division bat detector divides the frequency of an incoming bat call by a predetermined ratio that lowers its frequency and, unlike a heterodyne detector, picks up bat calls within a much broader range of frequencies. A time-expansion bat detector slows down the duration of a bat call so that much more detail about the acoustic structure of the call can be discerned.

Question 13: What is being done to protect bats and how can I help?

Answer: Organizations like Bat Conservation International (BCI) are helping to protect bats and bat habitats in many ways. BCI produces education materials about bats, supports conservation-relevant research at universities worldwide, and works with partners in industry as well as local, state, and federal agencies to protect bat habitats in forests, caves, rangelands, and wetlands. Diverse partners include the American Cave Conservation Association, the National Wildlife Federation, the Nature Conservancy, the U.S. Forest Service, Fish and Wildlife Service, Natural Resources Conservation Service, Bureau of Land Management, and the Department of Defense, to name a few. Such collaboration has led to protection of hundreds of critical caves and mines through the installation of bat-friendly gates and played a key role in recovery planning for endangered Indiana bats (*Myotis sodalis*). BCI also works with partners in Mexico to manage important caves used by bats that migrate there from the United States to overwinter, and research in collaboration with the USDA Natural Resources Conservation Service led to an essential publication that instructs landowners on how to provide safe drinking water for bats and other wildlife. You, too, can help protect bats by joining BCI and supporting this work.

Other conservation groups also do valuable work to protect bats, including the Bat Conservation Trust, Bat World Sanctuary, the Indiana State University Center for North American Bat Research, and the Lubee Bat Conservancy, to name a few (see appendix A).

There are many other things you can do to help bats. You can tell your friends and acquaintances about the important things bats do for us, so they also understand why bats should be protected. Tell people that they can safely evict bats from a building without killing them and direct them to bat exclusion information on the Internet. Don't disturb bats roosting in caves. Explain that many bats give birth to their babies in the late spring or summer and hibernate in the winter and that it is particularly important not to approach bats during those times. Help support efforts to carefully manage forests and wetlands, so they continue to provide good places for bats to roost. You can put up a bat house or help support the construction of other artificial roosts that are built for bats. Buy books and other educational materials about bats for yourself and your friends.

The more people understand about bats, the less likely they are to harm them. Add your voice to the growing number of people working to protect bats and help maintain a healthy environment.

APPENDIX A

Resources

Bat Conservation International: A nonprofit organization involved in bat education, research, and habitat conservation worldwide, with approximately 13,000 members. Publishes the quarterly *Bats* magazine.
P.O. Box 162603
Austin, Texas 78716
www.batcon.org

Bat Conservation Trust: A nonprofit organization involved in bat education, research, and habitat conservation in Europe and other parts of the world.
Unit 2, 15 Cloisters House
8 Battersea Park Road
London SW8-4BG
www.bats.org.uk

Bat World Sanctuary: A nonprofit organization with multiple satellite facilities throughout the United States that provides sanctuary for orphaned and injured bats as well as for bats that have been used in research. The organization provides educational programs for thousands of schoolchildren each year.
217 N. Oak
Mineral Wells, Texas 76067
www.batworld.org

Indiana State University (ISU) Bat Center: The university bat center conducts applied research on North American bats by collaborating with students and other scientists and makes the findings available to the scientific community and the public through technical and popular publications, teaching, and outreach programs.

Center for North American Bat Research and Conservation
Department of Ecology and Organismal Biology
Indiana State University
Terre Haute, Indiana 47809
www1.indstate.edu/biology/centers/bat.htm

Lubee Bat Conservancy: A nonprofit organization that focuses on research and conservation of fruit and nectar bats and their habitats worldwide. Also educates hundreds of people each year.
1309 N.W. 192nd Avenue
Gainesville, Florida 32609
www.lubee.org

North American Symposium on Bat Research (NASBR): Annual symposium that promotes and develops all aspects of the scientific study of bats, including conservation and public education.
www.nasbr.org

Speleobooks: Sells books, posters, periodicals, videos, and gift items related to speleology (the study of caves) for cavers and bat enthusiasts.
P.O. Box 10
Schoharie, New York 12157
www.speleobooks.com

APPENDIX B

Suggestions for Further Reading

Altringham, J. D. 1996. *Bats: Biology and Behaviour.* New York and Tokyo: Oxford University Press.

Crichton, E. G., and P. H. Krutzch, eds. 2000. *Reproductive Biology of Bats.* London and San Diego: Academic Press.

Fenton, M. B. 1992. *Bats.* New York: Facts On File.

Griffin, D. R. 1986. *Listening in the Dark: The Acoustic Orientation of Bats and Men.* Ithaca, N.Y.: Comstock Publishing.

Halfmann, J. 2004. *Red Bat at Sleepy Hollow Lane.* Norwalk, Conn.: Soundprint. (Ages 4–8)

Hall, L., and G. Richards. 2000. *Flying Foxes, Fruit and Blossom Bats of Australia.* Sydney, New South Wales: University of New South Wales Press Ltd.

Harvey, M. J., J. S. Altenbach, and T. L. Best. 1999. *Bats of the United States.* Little Rock: Arkansas Game and Fish Commission.

Hutson, T. 2000. *Bats.* Stillwater, Minn.: Voyageur Press.

Kunz, T. H., and M. B. Fenton, eds. 2003. *Bat Ecology.* Chicago and London: University of Chicago Press.

LaVal, R. K., and B. Rodriguez-H. 2002. *Murciélagos de Costa Rica (Costa Rica Bats).* Costa Rica: Instituto Nacional de Biodiversidad.

Lollar, A. *Bats in the Pantry.* Mineral Wells, Tex.: Bat World Sanctuary.

Lollar, A., and B. Schmidt-French. 2002. *Captive Care and Medical Reference for the Rehabilitation of Insectivorous Bats.* 2nd ed. Mineral Wells, Tex.: Bat World Publications.

Neuweiler, G. 2000. *The Biology of Bats.* New York: Oxford University Press.

Reid, F. A. 2006. *A Field Guide to Mammals of North America.* 4th ed. New York: Houghton Mifflin Company.

Richardson, P. 2002. *Bats.* Washington, D.C.: Smithsonian Institution Press.

Swanson, D. 1998. *Welcome to the World of Bats.* Canada: Walrus Books. (Ages 8–12)

Tallamy, D. 2007. *Bringing Nature Home: How Native Plants Sustain Wildlife in Our Gardens.* Portland, Oreg.: Timber Press.

Taylor, P. J. 2000. *Bats of Southern Africa.* Scottsville, South Africa: University of Natal Press.

Thomas, J. A., C. F. Moss, and M. Vater. 2004. *Echolocation in Bats and Dolphins.* Chicago and London: University of Chicago Press.

Tuttle, M. D. 2005. *America's Neighborhood Bats.* 2nd rev. ed. Austin, Tex.: University of Texas Press.

Whitaker, J. O., Jr. 1996. *National Audubon Society Field Guide to North American Mammals.* 2nd ed. New York: Alfred A. Knopf.

Whitaker, J. L., Jr., and W. J. Hamilton Jr. 1998. *Mammals of the Eastern United States.* Ithaca and London: Cornell University Press.

Find bat watching sites at http://www.tpwd.state.tx.us/publications/pwdpubs/media/pwd_bk_w7000_1411.pdf.

References

Chapter One: Bat Basics

Question 2: How are bats classified?

Kunz, T. H., and M. B. Fenton. 2003. *Bat Ecology*. Chicago: University of Chicago Press.

Reeder, D. M., et al. 2007. "Global Trends and Biases in New Mammal Species Discoveries." *Museum of Texas Tech University* 269:1–35.

Question 4: Where in the world are bats found?

Surlykke, A. 1986. "Moth Hearing on the Faroë Islands, an Area Without Bats." Physical Entomology 11:221–225.

Question 5: When did bats evolve?

Gunnell, G. F., et al. 2008. "New Bats (Mammalia: Chiroptera) from the Late Eocene and Early Oligocene, Fayum Depression, Egypt." *Journal of Vertebrate Paleontology* 28:1–11.

Jones, K. E., et al. 2005. "Bats, Clocks, and Rocks: Diversification Patterns in Chiroptera." *Evolution* 59:2243–2255.

Kriegs, J. O., et al. 2006. "Retroposed Elements as Archives for the Evolutionary History of Placental Mammals." *PloS Biology* 4:537–544.

Nikaido, M., et al. 2000. "Monophyletic Origin of the Order Chiroptera and Its Phylogenetic Position among Mammalia, as Inferred from the Complete Sequence of the Mitochondrial DNA of a Japanese Megabat, the Ryukyu Flying Fox (*Pteropus dasymallus*)." *Journal of Molecular Evolution* 51:318–328.

Nishihara, H., et al. 2006. "Pegasoferae, an Unexpected Mammalian Clade Revealed by Tracking Ancient Retroposon Insertions." *Proceedings of the National Academy of Sciences* 103:9929–9934.

Sears, K. E., et al. 2004. "Development of Bat Flight: Morphologic and Molecular Evolution of Bat Wing Digits." *Proceedings of the National Academy of Sciences* 103:6581–6586.

Simmons, N. B., et al. 2008. "Primitive Early Eocene Bat from Wyoming and the Evolution of Flight and Echolocation." *Nature* 451:818–821.

Speakman, J. 2008. "Evolutionary Biology: A First for Bats." *Nature* 451:774–775.

Speakman, J. R., et al. 1989. "The Energy Cost of Echolocation in Pipistrelle Bats (*Pipistrettus pipistrellus*)." *Journal of Comparative Physiology A: Neuroethology, Sensory, Neural, and Behavioral Physiology* 165:679–685.

Speakman, J. R., and P. A. Racey. 1991. "No Cost of Echolocation for Bats in Flight." *Nature* 350:421–423.

Question 7: How long do bats live?

Podlutsky, A. J., et al. 2005. "A New Field Record for Bat Longevity." *Journal of Gerontology: Biological Sciences* 60A:1366–1368.

Wilkinson, G. S., and J. M. South. 2002. "Life History, Ecology and Longevity in Bats." *Aging Cell* 1:124–131.

Question 8: Which is the biggest bat?

Kunz, T. H., and E. D. Pierson. 1994. "Introduction." In *Walker's Bats of the World,* R. M. Nowak, 1–46. Baltimore and London: Johns Hopkins University Press.

Mildenstein, T. L., et al. 2005. "Habitat Selection of Large Flying Foxes using Radio Telemetry: Targeting Conservation Efforts in Subic Bay, Philippines." *Biological Conservation* 126:93–102.

Nowak, R. M. 1994. *Walker's Bats of the World.* Baltimore and London: The Johns Hopkins University Press.

Stier, S. C., and T. L. Mildenstein. 2005. "Dietary Habits of the World's Largest Bats: The Philippine Flying Foxes, *Acerodon jubatus* and *Pteropus vampyrus lanensis*." *Journal of Mammalogy* 86:719–728.

Question 10: How far can bats fly?

Hill, J. E., and J. D. Smith. 1984. *Bats: A Natural History.* Austin: University of Texas Press.

Kunz, T. H., and M. B. Fenton. 2003. *Bat Ecology.* Chicago: University of Chicago Press.

Question 11: How fast do bats fly?

Hedenstrom, A., et al. 2007. "Bat Flight Generates Complex Aerodynamic Tracks." *Science* 316:894–897.

Hill, J. E., and J. D. Smith. 1984. *Bats: A Natural History.* Austin: University of Texas Press.

Kunz, T. H., and E. D. Pierson. 1994. "Introduction." In *Walker's Bats of the World,* R. M. Nowak, 1–46. Baltimore and London: The Johns Hopkins University Press.

Muijres, F. T., et al. 2008. "Leading–Edge Vortex Improves Lift in Slow-Flying Bats." *Science* 319:1250–1253.

Tuttle, M. D. 1994. "The Lives of Mexican Free-Tailed Bats." *Bats* 13: 6–14.

Question 12: How high do bats fly?

Capinera, J. L. "Featured Creatures: Tobacco Budworm." On the Web site University of Florida Institute of Food and Agricultural Sciences, Department of Entomology and Nematology. http://creatures.ifas.ufl.edu/field/tobacco_budworm.htm; accessed February 26, 2008.

Kunz, T. H., and M. B. Fenton. 2003. *Bat Ecology.* Chicago: University of Chicago Press.

McCracken, G. 1996. "Bats Aloft: A Study of High Altitude Feeding." *Bats* 14:7–10.

Tuttle, M. D. 1994. "The Lives of Mexican Free-Tailed Bats." *Bats* 13:6–14.

Question 13: Are bats intelligent?

Bohn, K. M., B. Schmidt-French, T. Ma, and G. D. Pollak. 2008. "Syllable Acoustics, Temporal Patterns and Call Composition Vary with Behavioral Context in Mexican Free-Tailed Bats." *Journal of the Acoustical Society of America* 24:1838–1848.

Boughman, J. W. 1998. "Voice Learning by Greater Spear-Nosed Bats." *Proceedings of the Royal Society B: Biological Sciences* 265:227–233.

Dunbar, R. I. 2003. "The Social Brain: Mind, Language, and Society in Evolutionary Perspective." *Annual Review of Anthropology* 32:163–181.

French, B. 1995. "The Song of the Mexican Free-Tail." *Bats* 13:7–9.

Jones, G., and M. W. Holderied. 2007. "Bat Echolocation Calls: Adaptation and Convergent Evolution." *Proceedings of the Royal Society B: Biological Sciences* 274:905–912.

Kerth, G., et al. "Group Decision Making in Fission-Fusion Societies: Evidence from Two Field Experiments in Bechstein's Bats." *Proceedings of the Royal Society B: Biological Sciences* 273:2785–2790.

King, B. J. 1986. "Extractive Foraging and the Evolution of Primate Intelligence." *Journal of Human Evolution* 1:361–372.

Lollar, A. 1995. "Notes on the Mating Behavior in a Captive Colony of Mexican Free-Tailed Bats." *Bat Research News* 36:5.

Lollar, A., and B. Schmidt-French. 2002. *Captive Care and Medical Reference for the Rehabilitation of Insectivorous Bats.* 2nd ed. Mineral Wells, Tex.: Bat World Publications.

Page, R. 2004. "Little Old Man: A Tale of Frogs, Fish and Amazing Memories." *Bats* 22:1–2.

Page, R., and M. Ryan. 2006. "Social Transmission of Novel Foraging Behavior in Bats: Frog Calls and Their Referents." *Current Biology* 16:1201–1205.

Ratcliffe, J. M., and H. M. ter Hofstede. 2005. "Roosts as Information Centres: Social Learning of Food Preferences in Bats." *Biology Letters* 1:72–74.

Rydell, J., and R. Arlettaz. 1994. "Low-Frequency Echolocation Enables the Bat *Tadarida tenotis* to Feed on Tympanate Insects." *Proceedings of the Royal Society B: Biological Sciences* 257:175–178.

Safi, K., and D.K.N. Dechmann. 2005. "Adaptation of Brain Regions to Habitat Complexity: A Comparative Analysis in Bats (Chiroptera)." *Proceedings of the Royal Society B: Biological Sciences* 272:179–186.

Safi, K., et al. 2005. "Bigger Is Not Always Better: When Brains Get Smaller." *Biology Letters* 1:283–286.

Singaravelan, N., and G. Marimuthu. 2007. "In Situ Feeding Tactics of Short-Nosed Fruit Bat (*Cynopterus sphinx*) on Mango Fruits: Evidence of Extractive Foraging in a Winged Mammal." *Journal of Ethology* 26:1–7.

Tuttle, M. D. 2005. *America's Neighborhood Bats.* 2nd rev. ed. Austin: University of Texas Press.

Question 14: Do bats really drink blood?

Bahlman, J. W., and D. A. Kelt. 2007. "Use of Olfaction during Prey Location by the Common Vampire Bat (*Desmodus rotundus*)." *Biotropica* 39:147–149.

Brown, D. E. 1994. *Vampiro: The Vampire Bat in Fact and Fantasy.* Silver City, N.Mex.: High-Lonesome Books.

Carter, G., et al. 2008. "Antiphonal Calling Allows Individual Discrimination in White-Winged Vampire Bats." *Animal Behaviour.* In press.

Carter, G., et al. 2006. "Avian Host DNA Isolated from the Feces of White-Winged Vampire Bats (*Diaemus youngi*)." *Acta Chiropterologica* 8:255–259.

Groger, U., and L. Wiegrebe. 2006. "Classification of Human Breathing Sounds by the Common Vampire Bat, *Desmodus rotundus.*" *BMC Biology* 4:4–18.

Chapter Two: Bat Bodies

Question 1: How are bats different from birds?

Hill, J. E., and J. D.Smith. 1984. *Bats: A Natural History*. Austin: University of Texas Press.

Question 2: Why do bats hang upside down?

Hill, J. E., and J. D. Smith. 1984. *Bats: A Natural History*. Austin: University of Texas Press.

Question 3: Do bats have teeth?

Neuweiler, G. 2000. *The Biology of Bats*. New York: Oxford University Press.

Question 4: Can bats walk?

Goodman, S. M., et al. 2007. "The Description of a New Species of *Myzopoda* (Myzopodidae: Chiroptera) from Western Madagascar." *Mammalian Biology* 72:65–81.
Riskin, D. K., and J. W. Hermanson. 2005. "Biomechanics: Independent Evolution of Running in Vampire Bats. *Nature:* 434:292.

Question 5: How fast do bats grow?

Hill, J. E., and J. D. Smith. 1984. *Bats: A Natural History*. Austin: University of Texas Press.
Kunz, T. H. and M. B. Fenton. 2003. *Bat Ecology*. Chicago: University of Chicago Press.
Kunz, T. H., and W. R. Hood. 2000. "Prenatal Care and Postnatal Growth in the Chiroptera." In *Reproductive Biology of Bats,* ed. E. G. Crichton and P. H. Krutzsch, 415–454. London: Academic Press.
Lollar, A., and B. Schmidt-French. 2002. *Captive Care and Medical Reference for the Rehabilitation of Insectivorous Bats*. 2nd ed. Mineral Wells, Tex.: Bat World Sanctuary.

Question 6: Are bats blind?

Fenton, M. B. 1992. *Bats*. New York: Facts On File.
Kunz, T. H., and M. B. Fenton. 2003. *Bat Ecology*. Chicago: University of Chicago Press.
Kunz, T. H., and E. D. Pierson. 1994. "Introduction." In *Walker's Bats of the World,* R. M. Nowak, 1–46. Baltimore and London: Johns Hopkins University Press.

Schmidt, U., et al. 1998. "On Some Unexpected Abilities of the Visual System in Phyllostomid Bats." *Bat Research News* 39:81–82.

Question 7: Why do bats have big ears?

Fenton, M. B. 1992. *Bats.* New York: Facts On File.

Hill, J. E., and J. D. Smith. 1992. *Bats: A Natural History.* Austin: University of Texas Press.

Kunz, T. H., and M. B. Fenton. 2003. *Bat Ecology.* Chicago: University of Chicago Press.

Kunz, T. H., and E. D. Pierson. 1994. "Introduction." In *Walker's Bats of the World,* R. M. Nowak, 1–46. Baltimore and London: The Johns Hopkins University Press.

Neuweiler, G. 2000. *The Biology of Bats.* New York: Oxford University Press.

Chapter Three: Bat Life

Question 1: What do bats eat?

Fenton, M. B. 1992. *Bats.* New York: Facts On File.

Freeman, P. W., and C. A. Lemen. 2007. "Using Scissors to Quantify Hardness of Insects: Do Bats Select for Size or Hardness?" *Journal of Zoology* 271:469–476.

French, B. 1997. "False Vampires and Other Carnivores." *Bats* 15:11–14.

Kalka, M. B., et al. 2008. "Bats Limit Arthropods and Herbivory in a Tropical Forest." *Science* 320:71.

Kunz, T. H., and M. B. Fenton. 2003. *Bat Ecology.* Chicago: University of Chicago Press.

"Mud, Mud, Glorious Mud?" on the Web site ScienceDaily. www.sciencedaily.com/releases/2007/11/071117103531.htm; accessed February 29, 2008.

Novacek, M. 2007. *Terra.* New York: Farrar, Straus and Giroux.

Nowak, R. M. 1994. *Walker's Bats of the World.* Baltimore and London: Johns Hopkins University Press.

Tuttle, M. D. 2005. *America's Neighborhood Bats.* 2nd rev. ed. Austin: University of Texas Press.

Voigt, C. 2008. "Nutrition or Detoxification: Why Bats Visit Mineral Licks of the Amazonian Rainforest." PLoS ONE 3:e2011.

Whitaker, J. L., Jr., and W. J. Hamilton Jr. 1998. *Mammals of the Eastern United States.* Ithaca and London: Cornell University Press.

Williams-Guillen, K., et al. 2008. "Bats Limit Insects in a Neotropical Agroforestry System." *Science* 320:70.

Question 2: Where do bats live?

Fenton, M. B. 1992. *Bats*. New York: Facts On File.

French, B. 1999. "Where the Bats Are—Part I: Plants and Trees." *Bats* 17:10–13.

———. 1999. "Where the Bats Are—Part II: Other Animals' Shelters." *Bats* 17:14–16.

Kunz, T. H., and M. B. Fenton. 2003. *Bat Ecology*. Chicago: University of Chicago Press.

Nowak, R. M. 1994. *Walker's Bats of the World*. Boston and London: Johns Hopkins University Press.

Tuttle, M. D. 2005. *America's Neighborhood Bats*. 2nd rev. ed. Austin: University of Texas Press.

———. 2000. "Where the Bats Are—Part III: Caves, Cliffs, and Rock Crevices." *Bats* 18:6–11.

Question 3: Why do bats like caves?

French, B. 1999. "Where the Bats Are—Part I: Plants and Trees." *Bats* 17:10–13.

———. 1999. "Where the Bats Are—Part II: Other Animals' Shelters." *Bats* 17:14–16.

Kunz, T. H., and M. B. Fenton. 2003. *Bat Ecology*. Chicago: University of Chicago Press.

Tuttle, M. D. 2000. "Where the Bats Are—Part III: Caves, Cliffs, and Rock Crevices." *Bats* 18:6–11.

Question 4: Do bats only fly at night?

Easterla, D. A. 1972. "A Diurnal Colony of Big Freetail Bats, *Tadarida macrotis* (Gray), in Chihuahua, Mexico." *American Midland Naturalist* 88:468–470.

Elmore, L.W., et al. 2004."Selection of Diurnal Roosts by Red Bats (*Lasiurus borealis*) in an Intensively Managed Pine Forest in Mississippi." *Forest Ecology and Management* 199:11–20.

Fullard, J. H. 2000." Day-Flying Butterflies Remain Day-Flying in a Polynesian, Bat-Free Habitat." *Proceedings of the Royal Society: Biological Sciences* 1267:2295–2300.

Kunz, T. H., and E. D. Pierson. 1994. "Introduction." In *Walker's Bats of the World,* R. M. Nowak, 1–46. Baltimore and London: The Johns Hopkins University Press.

Myers, N. 1979. *The Sinking Ark: A New Look at the Problem of Disappearing Species*. New York: Readers Digest Young Families.

Sewall, B. J. 2003. "Giant Bats Face a Shrinking Forest: Conserving Livingstone's Flying Foxes." *Bats* 21:8–11.

Simmons, N. B., et al. 2008. "Primitive Early Eocene Bat from Wyoming and the Evolution of Flight and Echolocation." *Nature* 451:818–821.

Stein, C. 1997. "A National Park for Bats." *Bats* 15:12–13.

Thompson, S. C., and J. R. Speakman. 1999. "Absorption of Visible Spectrum Radiation by the Wing Membranes of Living Pteropodid Bats." *Journal of Comparative Physiology B: Biochemical, Systemic, and Environmental Physiology* 169:187–194.

Tuttle, M. D. 1997. "On the Cover." *Bats* 15:1–2.

Question 5: What do bats do during the day?

French, B. 1995. "The Song of the Mexican Free-Tail." *Bats* 13:7–9.

French, B., and A. Lollar. 2000. "Communication among Mexican Free-Tailed Bats." *Bats* 18:1–6.

Hill, J. E., and J. D. Smith. 1984. *Bats: A Natural History*. Austin: University of Texas Press.

Kunz, T. H., and M. B. Fenton. 2003. *Bat Ecology*. Chicago: University of Chicago Press.

Kunz, T. H., and E. D. Pierson. 1994. "Introduction." In *Walker's Bats of the World,* R. M. Nowak, 1–46. Baltimore and London: Johns Hopkins University Press.

Locke, R. 2004. "Bat Talk: Do Bats Possess Language?" *Bats* 22:1–6.

Lollar, A. 1995. "Notes on the Mating Behavior of a Captive Colony of *Tadarida brasiliensis*." *Bat Research News* 36:5.

Lollar, A., and B. Schmidt-French. 2002. *Captive Care and Medical Reference for the Rehabilitation of Insectivorous Bats*. 2nd ed. Mineral Wells, Tex.: Bat World Publication.

Question 6: Do all bats live in groups?

Fenton, M. B. 1992. *Bats*. New York: Facts On File.

Graham, G. L. 1988. "Interspecific Associations among Peruvian Bats at Diurnal Roosts and Roost Sites." *Journal of Mammology* 69:711–720.

Hill, J. E., and J. D. Smith. 1984. *Bats: A Natural History*. Austin: University of Texas Press.

Kunz, T. H., and M. B. Fenton. 2003. *Bat Ecology*. Chicago: University of Chicago Press.

Kunz, T. H., and E. D. Pierson. 1994. "Introduction." In *Walker's Bats of the World,* R. M. Nowak, 1–46. Baltimore and London: Johns Hopkins University Press.

Tuttle, M. D. 2005. *America's Neighborhood Bats.* 2nd rev. ed. Austin: University of Texas Press.

Question 7: What is bat guano?

Bernath, R. F., and T. H. Kunz. 1981. "Structure and Dynamics of Arthopod Communities in Bat Guano Deposits in Buildings." *Canadian Journal of Zoology* 59:260–270.

Fenolio, D. 2005. "Population Ecology and Behavior of the Ozark Blind Cave Salamander, *Eurycea spelaea.*" Paper presented at the annual meeting of the Society of Integrative and Comparative Biology, January 4–8, San Diego, California.

Keleher, S. 1996. "Guano: Bats' Gift to Gardeners." *Bats* 14:15–17.

Ritzi, C. M., and C. M. Ritzi. 2001. "The Arthropod Community in Bat Guano from an Abandoned Building in Presidio County, Texas." *Texas Journal of Science* 53:79–82.

Steele, D. B. 1989. "Bats, Bacteria and Biotechnology." *Bats* 7:3–4.

Tuttle, M. D. 1994. "The Lives of Mexican Free-Tailed Bats." *Bats* 12:6–14.

———. 2005. *America's Neighborhood Bats.* 2nd rev. ed. Austin: University of Texas Press.

Webster, J., and J. O. Whitaker Jr. 2005. "Studies of Guano Communities in Big Brown Bat Colonies in Indiana Neighboring Illinois Counties." *Northeastern Naturalist* 12 (2):221–232.

Whitaker, J. O., Jr., P. Clem, and J. R. Munsee. 1991. "Trophic Structure of the Community of the Guano in the Brazilian Free-Tailed Bat (*Tadarida mexicana*) in Texas." *American Midland Naturalist* 126: 392–398.

Question 8: Do bats migrate?

Altringham, J. D. 1996. *Bats: Biology and Behaviour.* New York: Oxford University Press.

Cryan, P. M. 2003. "Seasonal Distribution of Migratory Tree Bats in North America." *Journal of Mammalogy* 84:579–593.

Cryan, P. M., and B. O. Wolf. 2003. "Sex Differences in the Thermoregulation and Evaporative Water Loss of a Heterothermic Bat, *Lasiurus cinereus,* during its Spring Migration." *Journal of Experimental Biology* 206:3381–3390.

Cryan, P. M., et al. 2004. "Stable Hydrogen Isotope Analysis of Bat Hair as Evidence for Seasonal Molt and Long-Distance Migration." *Journal of Mammalogy* 85:995–1001.

Fenton, M. B. 1992. *Bats.* New York: Facts On File.

Hill, J. E., and J. D. Smith. 1984. *Bats: A Natural History.* Austin: University of Texas Press.

Holland, R. A., et al. 2008. "Bats Use Magnetite to Detect the Earth's Magnetic Field." *PLoS ONE.* 3:e1676.

Kunz, T. H., and M. B. Fenton. 2003. *Bat Ecology.* Chicago: University of Chicago Press.

Neuweiler, G. 2000. *The Biology of Bats.* New York: Oxford University Press.

Tuttle, M. D. 2005. *America's Neighborhood Bats.* 2nd rev. ed. Austin: University of Texas Press.

Question 9: Do bats hibernate?

Altringham, J. D. 1996. *Bats: Biology and Behaviour.* New York: Oxford University Press.

Fenton, M. B. 1992. *Bats.* New York: Facts On File.

Hill, J. E., and J. D. Smith. 1984. *Bats: A Natural History.* Austin: University of Texas Press.

Kunz, T. H., and M. B. Fenton. 2003. *Bat Ecology.* Chicago: University of Chicago Press.

Lollar A., and B. Schmidt-French. 2002. *Captive Care and Medial Reference for the Rehabilitation of Insectivorous Bats.* 2nd ed. Mineral Wells, Tex.: Bat World Sanctuary.

Neuweiler, G. 2000. *The Biology of Bats.* New York: Oxford University Press.

Chapter Four: Bat Behavior

Question 1: How does echolocation work?

Altringham, J. 1996. *Bats, Biology and Behaviour.* London: Oxford University Press.

Bohn, K. M., et al. 2006. "Correlated Evolution between Hearing Sensitivity and Social Calls in Bats." *Biology Letters* 2:561–564.

Fenton, M. B. 1992. *Bats.* New York: Facts On File.

Gillam, E. H. 2007. "Eavesdropping by Bats on the Feeding Buzzes of Conspecifics." *Canadian Journal of Zoology* 85:795–801.

Gillam, E. H., et al. 2007. "Rapid Jamming Avoidance in Biosonar." *Proceedings of the Royal Society B: Biological Sciences* 274:651–660.

Gillam, E. H., and G. F. McCracken. 2007. "Variability in the Echolocation of Brazilian Free-Tailed Bats, *Tadarida brasiliensis*: Effects of Geography and Local Acoustic Environment." *Animal Behaviour* 74:277–286.

Griffin, D. R. 1958. *Listening in the Dark: The Acoustic Orientation of Bats and Men.* New York: Dover Publications.

Jones, G., and R. D. Ransome. 1993. "Echolocation Calls of Bats Are Influenced by Maternal Effects and Change over a Lifetime." *Proceedings of the Royal Society B: Biological Sciences* 252:125–128.

Kunz, T. H., and M. B. Fenton. 2003. *Bat Ecology*. Chicago: University of Chicago Press.

Ma, J., et al. 2006. "Vocal Communication in Adult Greater Horseshoe Bats, *Rhinolophus errumequinum*." *Journal of Comparative Physiology A: Neuroethology, Sensory, Neural, and Behavioral Physiology* 192:535–550.

Pollak, G., et al. 1977. "Echo-Detecting Characteristics of Neurons in Inferior Colliculus of Unanesthetized Bats." *Science* 196:675–678.

Tuttle, M. D. 2005. *America's Neighborhood Bats*. 2nd rev ed. Austin: University of Texas Press.

Question 2: How do bats navigate in the dark?

Altringham, J., et al. 1996. *Bats, Biology and Behaviour*. London: Oxford University Press.

Griffin, D. R. 1958. *Listening in the Dark: The Acoustic Orientation of Bats and Men*. New York: Dover Publications.

Holland, R. A., et al. 2006. "Navigation: Bat Orientation Using Earth's Magnetic Field." *Nature* 444:702.

Holland, R. A., et al. 2008. "Bats Use Magnetite to Detect the Earth's Magnetic Field." *PLoS ONE* 3:1–6.

Neuweiler, G. 2000. *The Biology of Bats*. New York: Oxford University Press.

Wang, Y., et al. 2007. "Bats Respond to Polarity of a Magnetic Field." *Proceedings of the Royal Society B: Biological Sciences* 274:2901–2905.

Waters, C. A., and J. G. Wong. 2007. "The Allocation of Energy to Echolocation Pulses Produced by Soprano Pipistrelles (*Pipistrellus pygmaeus*) during the Wingbeat Cycle." *Journal of the Acoustical Society of America* 21:2990–3000.

Question 3: Do all bats use echolocation to find food?

Fenton, M. B. 1992. *Bats*. New York: Facts On File.

Hahn, W. L. 1908. "Some Habits and Sensory Adaptations of Cave-Inhabiting Bats." *Biological Bulletin* 3:135–164.

Holland, R. A., et al. 2005. "Sensory Systems and Spatial Memory in the Fruit Bat *Rousettus aegyptiacus*." *Ethology* 111:715–725.

Kunz, T. H., and M. B. Fenton. 2003. *Bat Ecology*. Chicago: University of Chicago Press.

Kunz, T. H., and E. D. Pierson. 1994. "Introduction." In *Walker's Bats of the World*, R. M. Nowak, 1–46. Baltimore and London: Johns Hopkins University Press.

Mueller, H. C., and N. S. Mueller. 1979. "Sensory Basis for Spatial Memory in Bats." Journal of Mammalogy 60:198–201.

Thiele, J., and Y. Winter. 2005. "Hierarchical Strategy for Relocating Food Targets in Flower Bats: Spatial Memory versus Cue-Directed Search." *Animal Behaviour* 69:315–327.

Question 4: Are bats the only animals that use echolocation?

Buchler, E. R. 1976. "Experimental Demonstration of Echolocation by the Wandering Shrew (*Sorex vagrans*)." *Animal Behaviour* 24:858–873.

Konisi, M., and E. I. Knudsen. 1979. "The Oilbird: Hearing and Echolocation." *Science* 27:425–427.

Price, J. J., et al. 2004. "The Evolution of Echolocation in Swiftlets." *Journal of Avian Biology* 35:135–143.

Thomas, J. A. 2004. *Echolocation in Bats and Dolphins.* Chicago: University of Chicago Press.

Thomasi, T. E. 1979. "Echolocation by the Short-Tailed Shrew *Blarina brevicauda.*" *Journal of Mammalogy* 60:751–759.

Question 5: How does a bat's prey defend itself?

Barber, J. R., and W. E. Conner. 2006. "Tiger Moth Responses to a Simulated Bat Attack: Timing and Duty Cycle." *Journal of Experimental Biology* 209:2637–2650.

———. 2007. "Acoustic Mimicry in a Predator-Prey Interaction." *Proceedings of the National Academy of Sciences* 104:9331–9334.

Fullard, J. H., et al. 1994. "Jamming Bat Echolocation: The Dogbane Tiger Moth *Cycnia tenera* Times Its Clicks to the Terminal Attack Calls of the Big Brown Bat *Eptesicus fuscus.*" *Journal of Experimental Biology* 194:285–298.

Hoy, R. R. 1992. "The Evolution of Hearing in Insects as an Adaptation to Predation by Bats." In *The Evolutionary Biology of Hearing,* ed. D. B. Webster, R. R. Fay, and A. N. Popper, 115–129. New York: Springer.

Hristov, N., and W. E. Connor. 2005. "Effectiveness of Tiger Moth (Lepidoptera, Arctiidae) Chemical Defenses against an Insectivorous Bat (*Eptesicus fuscus*)." *Chemoecology* 15:105–113.

———. 2005. "Sound Strategies: Acoustic Aposematism in the Bat Tiger Moth Arms Race." *Naturwissenschaften* 92:164–169.

Miller, L. A., and A. Surlykke. 2001. "How Some Insects Detect and Avoid Being Eaten by Bats: Tactics and Countertactics of Prey and Predator." *BioScience* 51:570–581.

Rydell, J. 1998. "Bat Defense in Lekking Ghost Swifts (*Hepialus humuli*), a Moth Without Ultrasonic Hearing." *Proceedings of the Royal Society B: Biological Sciences* 265:1373–1376.

Question 6: How do bats communicate?

Altringham, J. D., and M. B. Fenton. 2003. "Sensory Ecology and Communication in the Chiroptera." In *Bat Ecology*, ed. T. H. Kunz and M. B. Fenton, 90–127. Chicago: University of Chicago Press.

Balcombe, J. P., and G. F. McCracken. 1992. "Vocal Recognition in Mexican Free-Tailed Bats: Do Pups Recognize Mothers?" *Animal Behaviour* 43:79–87.

Behr, O., and O. von Helversen. 2004. "Bat Serenades—Complex Courtship Songs of the Sac-Winged Bat (*Saccopteryx bilineata*)." *Behavioral Ecology and Sociobiology* 56:106–115.

Behr, O., et al. 2006. "Territorial Songs Indicate Male Quality in the Sac-Winged Bat *Saccopteryx bilineata* (Chiroptera, Emballonuridae)." *Behavioral Ecology* 17:810–817.

Bohn, K. M., et al. 2007. "Discrimination of Infant Isolation Calls by Greater Spear-Nosed Bats, *Phyllostomus hastatus*." *Animal Behaviour* 73:423–432.

Bohn, K., B. Schmidt-French, S. Ma, and G. Pollak. 2008. "Syllable Acoustics, Temporal Patterns and Call Composition Vary with Behavioral Context in Mexican Free-Tailed Bats." *Journal of the Acoustical Society of America* 124:1838–1848.

Carter, G., et al. 2007. *Vocal Communication in White-Winged Vampires (Diaemus youngi)*. Paper presented at the fourteenth International Bat Research Conference, Merida, Mexico.

Carter, G., et al. 2008. "Antiphonal Calling Allows Individual Discrimination in White-Winged Vampire Bats." *Animal Behaviour*. In press.

Davidson, S. M., and G. S. Wilkinson. 2002. "Geographic and Individual Variation in Vocalizations by Male *Saccopteryx bilineata* (Chiroptera: Emballonuridae)." *Journal of Mammalogy* 83:526–535.

Fenton, M. B. 1992. *Bats*. New York: Facts On File.

———. 2003. "Eavesdropping on the Echolocation and Social Calls of Bats." *Mammal Review* 33:193–204.

French, B. 2003. "Bat Smells: How Aromas Affect Bat Behavior." *Bats* 21:12–13.

French, B., and A. Lollar. 2000. "Letters to the Editor." *Bat Research News* 41:26–28.

Kanwal, J. S., et al. 1994. "Analysis of Acoustic Elements and Syntax in Communication Sounds Emitted by Moustached Bats." *Journal of the Acoustical Society of America* 96:1229–1254.

Kazial, K. A., and W. M. Masters. 2004. "Female Big Brown Bats, *Eptesicus fuscus*, Recognize Sex from a Caller's Echolocation Signals." *Animal Behaviour* 67:855–863.

Knörnschild, M., et al. 2006. Babbling Behavior in the Sac-Winged Bat (*Saccopteryx bilineata*). *Naturwissenschaften* 93:451–454.

Kunz, T. H., and M. B. Fenton. 2003. *Bat Ecology*. Chicago: University of Chicago Press.

Locke, R. 2004. "Bat Talk." *Bats* 22:1–6.

Lollar, A., and B. Schmidt-French. 2002. *Captive Care and Medical Reference for the Rehabilitation of Insectivorous Bats.* 2nd ed. Mineral Wells, Tex.: Bat World Publication.

Ma, J., et al. 2006. "Vocal Communication in Adult Greater Horseshoe Bats, *Rhinolophus ferrumequinum.*" *Journal of Comparative Physiology A: Neuroethology, Sensory, Neural, and Behavioral Physiology* 192:535–550.

Masters, W. M., et al. 1995. "Sonar Signals of Big Brown Bats, *Eptesicus Fuscus,* Contain Information about Individual Identity, Age and Family Affiliation." *Animal Behaviour* 50:1243–1260.

Melendez, K. V., et al. 2006. "Classification of Communication Signals of the Little Brown Bat." *Journal of the Acoustical Society of America* 120:1095–1102.

Milius, S. 2000. "Male Bats Primp Daily for Odor Display." *Science News* 157:7.

Nielsen, L. T., D. K. Eaton, D. W. Wright, and B. A. French. 2006. "Characteristic Odors of *Tadarida brasiliensis mexicana* Chiroptera: Molossidae." *Journal of Cave and Karst Studies* 68:27–31.

Pearl, D. L., and M. B. Fenton. 1996. "Can Echolocation Calls Provide Information about Group Identity in the Little Brown Bat (*Myotis lucifugus*)." *Canadian Journal of Zoology* 74:2184–2192.

Pfalzer, G., and J. Kusch. 2003. "Structure and Variability of Bat Social Calls: Implications for Specificity and Individual Recognition." *Journal of Zoology* 261:21–33.

Pollack, G., et al. 1977. "Echo-Detecting Characteristics of Neurons in Inferior Colliculus of Unanesthetized Bats." *Science* 196:675–678.

Schmidt-French, B., E. Gillam, and M. B. Fenton. 2006. "Vocalizations Emitted during Mother/Young Interactions by Captive Eastern Red Bats *Lasiurus borealis* (Chiroptera: Vespertilionidae)." *Acta Chiropterological* 8:477–484.

Wilkinson, G. S., and J. W. Boughman. 1998. "Social Calls Coordinate Foraging in Greater Spear-Nosed Bats." *Animal Behaviour* 55: 337–350.

Question 7: How do bats manage extreme heat?

Kunz, T. H., and M. B. Fenton. 2003. *Bat Ecology*. Chicago: University of Chicago Press.

Welbergen, J. A., et al. 2008. "Climate Change and the Effects of Temperature Extremes on Australian Flying-Foxes." *Proceedings of the Royal Society B: Biological Sciences* 275:419–425.

Question 8: How do bats manage extreme cold?

Hill, J. E., and J. D. Smith. 1984. *Bats: A Natural History*. Austin: University of Texas Press.

Kunz, T. H., and M. B. Fenton. 2003. *Bat Ecology*. Chicago: University of Chicago Press.

Question 9: Can bats swim?

Lollar, A., and B. Schmidt-French. 2002. *Captive Care and Medial Reference for the Rehabilitation of Insectivorous Bats*. 2nd ed. Mineral Wells, Tex.: Bat World Sanctuary.

Nowak, R. M. 1994. *Walker's Bats of the World*. Baltimore and London: Johns Hopkins University Press.

Chapter Five: Bat Love

Question 1: How does a bat attract a mate?

Fenton, M. B. 1992. *Bats*. New York: Facts On File.

Kunz, T. H., and M. B. Fenton. 2003. *Bat Ecology*. Chicago: University of Chicago Press.

McCracken, G. F., and G. S. Wilkinson. 2000. "Bat Mating Systems." In *Reproductive Biology of Bats,* ed. E. G. Crichton and P. H. Krutzsch, 321–362. London and San Diego: Academic Press.

Question 2: Are bats monogamous?

McCracken, G. F., and G. S. Wilkinson. 2000. "Bat Mating Systems." In *Reproductive Biology of Bats,* ed. E. G. Crichton and P. H. Krutzsch, 321–362. London and San Diego: Academic Press.

Question 3: How do bats reproduce?

Kunz, T. H., and M. B. Fenton. 2003. *Bat Ecology*. Chicago: University of Chicago Press.

Kunz, T. H., et al. 1994. "Alloparental Care: Helper-Assisted Birth in the Rodrigues Fruit Bat *Pteropus rodricensis* (Chiroptera: Pteropodidae)." *Journal of Zoology* 232:691–700.

McCracken, G. F., and G. S. Wilkinson. 2000. "Bat Mating Systems." In *Reproductive Biology of Bats,* ed. E. G. Crichton and P. H. Krutzsch, 321–362. London and San Diego: Academic Press.

Pitnick, S., et al. 2006. "Mating System and Brain Size in Bats." *Proceedings of the Royal Society B: Biological Sciences* 273:719–724.

Wilkinson, G. S., and G. F. McCracken. 2003. "Bats and Balls: Sexual Selection and Sperm Competition in the Chiroptera." In *Bat Ecology,* ed. T. H. Kunz and M. B. Fenton, 128–155. Chicago and London: University of Chicago Press.

Question 4: How many pups are in a litter?

Crichton, E. G., and P. H. Krutzsch. 2000. *Reproductive Biology of Bats.* Cambridge: Cambridge University Press.

Fenton, M. B. 1992. *Bats.* New York: Facts On File.

Hill, J. E., and J. D. Smith. 1984. *Bats: A Natural History.* Austin: University of Texas Press.

Kunz, T. H., and M. B. Fenton. 2003. *Bat Ecology.* Chicago: University of Chicago Press.

Kunz, T. H., and W. E. Pierson. 1994. " Introduction." In *Walker's Bats of the World,* R. M. Nowak, 1–46. Baltimore and London: Johns Hopkins University Press.

Tuttle, M. D. 2005. *America's Neighborhood Bats.* 2nd rev. ed. Austin: University of Texas Press.

Question 5: Do bat mothers take care of their young?

Bohn, K. M., et al. 2006. "Correlated Evolution between Hearing Sensitivity and Social Calls in Bats." *Biology Letters* 2:561–564.

French, B. A., and J. O. Whitaker Jr. 2002. "Helping Orphans Survive." *Bats* 20:1–3.

Kunz, T. H., and M. B. Fenton. 2003. *Bat Ecology.* Chicago: University of Chicago Press.

Kunz, T. H., and W. R. Hood. 2000. "Prenatal Care and Postnatal Growth in the Chiroptera." In *Reproductive Biology of Bats,* ed. E. G. Crichton and P. H. Krutzsch, 415–454. London: Academic Press.

Schmidt-French, B., and J. O. Whitaker Jr. 2005. "Acquisition of Foraging Behavior and Insect Preferences by Naïve Juvenile Red Bats (*Lasiurus boralis*)." *Acta Chiropterologica* 7:314–318.

Question 6: Do bat fathers take care of their offspring?

Kunz, T. H., and W. R. Hood. 2000. "Prenatal Care and Postnatal Growth in the Chiroptera." In *Reproductive Biology of Bats,* ed. E. G. Crichton and P. H. Krutzsch, 415–454. London: Academic Press.

Question 7: How long does it take before newborn bats can fly?

Kunz, T. H., and M. B. Fenton. 2003. *Bat Ecology.* Chicago: University of Chicago Press.

Kunz, T. H., and W. R. Hood. 2000. "Prenatal Care and Postnatal Growth in the Chiroptera." In *Reproductive Biology of Bats,* ed. E. G. Crichton and P. H. Krutzsch, 415–454. London: Academic Press.

Kunz, T. H., and E. D. Pierson. 1994. "Introduction." In *Walker's Bats of the World,* R. M. Nowak, 1–46. Baltimore and London: Johns Hopkins University Press.

Chapter Six: Dangers and Defenses

Question 1: Are bats aggressive?

Lollar, A., and B. Schmidt-French. 2002. *Captive Care and Medial Reference for the Rehabilitation of Insectivorous Bats.* 2nd ed. Mineral Wells, Tex.: Bat World Sanctuary.

Tuttle, M. D. 2005. *America's Neighborhood Bats.* 2nd rev ed. Austin, Tex.: University of Texas Press.

Question 2: Do bats bite people?

Constantine, D. G. 1988. "Health Precautions for Bat Researchers." In *Ecological and Behavioral Methods for the Study of Bats,* ed. T. H. Kunz, 491–526. Washington, D.C.: Smithsonian Institute Press.

Lollar, A., and B. Schmidt-French. 2002. *Captive Care and Medial Reference for the Rehabilitation of Insectivorous Bats.* 2nd ed. Mineral Wells, Tex.: Bat World Sanctuary.

Question 3: Do all bats have rabies?

Blanton, J. D., et al. 2005. "Bats and Cats: Rabies Postexposure Prophylaxis." *Emerging Infectious Diseases* 11:1921–1927.

Brass, D. 1994. *Rabies in Bats: Natural History and Public Health Implications.* Ridgefield, Conn.: Livia Press.

Constantine, D. G. 1988. "Health Precautions for Bat Researchers." In *Ecological and Behavioral Methods for the Study of Bats,* ed. T. H. Kunz, 491–526. Washington, D.C.: Smithsonian Institute Press.

Recommendations of the Immunization Practices Advisory Committee. 1999. U. S. Department of Health and Human Services, Public Health Service, Center for Disease Control, National Center for Infectious Diseases, Division of Bacterial and Mycotic Diseases.

Recuenco, S., et al. 2007. "Potential Cost Savings with Terrestrial Rabies Control." *BMC Public Health* 7:47.

Schmidt-French, B., R. J. Rudd., and C. V. Trimarchi. 2006. "Frequency of Rabies Infections and Clinical Signs in a Wildlife Rehabilitation Sample of Bats in Central Texas, 1997–2001." *Journal of Wildlife Rehabilitation and Medicine* 28:5–8.

Schwiff, S. A., et al. 2007. "Direct and Indirect Costs of Rabies Exposure: A Retrospective Study in Southern California (1998–2002)." *Journal of Wildlife Diseases* 43:251–257.

Turmelle, A. S., et al. 2008. "Response to Vaccination with a Commercial Inactivated Rabies Vaccine in a Captive Colony of Brazilian Free-Tailed Bats (*Tadarida brasiliensis*)." Submitted to *Journal of Zoo and Wildlife Medicine.*

Whitaker, J. L., Jr., and W. J. Hamilton Jr. 1998. *Mammals of the Eastern United States.* Ithaca and London: Cornell University Press.

Question 4: Can people get diseases from bats?

Choi, C. "Bats Might Be Origin of SARS." On the Web site The Scientist. http://www.the-scientist.com/article/print/22783/; accessed April 17, 2008.

Hsu, V. P., et al. "Nipah Virus Encephalitis Reemergence, Bangladesh." On the Web site CDC: Emerging Infectious Diseases. http://www.cdc.gov/ncidod/EID/vol10no12/04-0701.htm; accessed April 17, 2008.

Li, W., et al. 2005. "Bats Are Natural Reservoirs of SARS-Like Coronaviruses." *Science* 310:676–679.

Mohd Yob, J., et al. "Nipah Virus Infection in Bats (Order Chiroptera) in Peninsular Malaysia." On the Web site CDC: Emerging Infectious Diseases. http://www.cdc.gov/ncidod/EID/vol7no3/yob.htm; accessed April 17, 2008.

Normile, D. 2005. "Researchers Tie Deadly SARS Virus to Bats." *Science* 309:2154–2155.

Quammen, D. 2007. "Deadly Contact." On the Web site National Geographic. http://magma.nationalgeographic.com/ngm/2007-10/infectious-animals/quammen-text.html; accessed April 17, 2008.

"Signs of Ebola Virus Found in Bats." On the Web site Center for Infectious Disease Research and Policy (CIDRAP). http://www.cidrap.umn.edu/cidrap/content/bt/vhf/news/nov3005bats.html; accessed April 22, 2008.

Question 5: What is White Nose Syndrome?

Blehert, D. S., et al. 2008. "Bat White-Nose Syndrome: An Emerging Fungal Pathogen?" *Science Express.* Published online October 30. www.sciencemag.org/cgi/content/abstract/1163874; accessed November 9, 2008.

Clark, D. R., Jr., and R. F. Shore. 2001. "Chiroptera." In *Ecotoxicology of Wild Mammals,* ed. R. F. Shore and B. A. Rattner, 159–214. New York: John Wiley and Sons.

Question 6: Do bats have enemies?

Barclay, R.M.R., et al. 1982. "Screech Owl, *Otus asio,* Attempting to Capture Little Brown Bats, *Myotis lucifugus,* at a Colony." *Canadian Field-Naturalist* 96:205–206.

Cary, D. L., et al. 1981. "An Observation of Snake Predation on a Bat." *Kansas Academy of Sciences: Transactions* 84:223–224.

Catto, C. 1994. "The Early Bat Gets Eaten by the Hawk." *Bat News* 32:3.

Claire, W., and M. A. Ports. 1981. "An Adaptive Method of Predation by the Great Horned Owl on Mexican Free-Tailed Bats." *Southwestern Naturalist* 26:69–70.

Constantine, D. G. 1948. "Great Bat Colonies Attract Predators." *Bulletin of the National Speleological Society* 10:100.

Fenton, M. B. 1992. *Bats.* New York: Facts On File.

Long, C. F. 1971. "Common Grackles Prey on Big Brown Bat." *Wilson Bulletin* 83:196.

"Make Your Garden Safe for Bats." On the Web site Bat Conserva-tion Trust. http://www.bats.org.uk/helpline/documents/Catastrophe-newlogopdf.pdf; accessed March 24, 2008.

McCracken, G. F., et al. 1986. "Raccoons Catch Mexican Free-Tailed Bats on the Wing." *Bat Research News* 27:21–22.

McFarlane, D. A., and K. L. Garrett. 1989. "The Prey of Common Barn Owls *Tyto alba* in Dry Limestone Scrub Forest of Southern Jamaica West Indies." *Caribbean Journal of Science* 35:21–23.

Munson, P. J., and J. H. Keith. 1984. "Prehistoric Raccoon Predation on Hibernating *Myotis,* Wyandotte Cave, Indiana." *Journal of Mammalogy* 65:152–155.

Ruprecht, A. L. 1979. "Bats (Chiroptera) as Constituents of the Food of Barn Owls *Tyto alba* in Poland." *Ibis* 121:489–494.

Sparks, D. W., et al. 2000. "Vertebrate Predators on Bats in North America North of Mexico." In *Reflections of a Naturalist: Papers Honor-*

ing Professor Eugene E. Fleharty, Fort Hayes Studies Special Issue, ed. J. R. Choate, 1:229–241.

Taylor, L. L., and S. M. Lehman. 1997. "Predation on an Evening Bat (*Nycticeius* sp.) by Squirrel Monkeys (*Saimiri sciureus*) in South Florida." *Anthropological Sciences* 60:112–117.

Van der Merwe, M. 1980. "Importance of *Miniopterus schreibersi natalensis* in the Diet of Barn Owls." *South African Journal of Wildlife Research* 10:15–17.

Vaughan, C. 1982. "Barn Owl Food." *Brenesia* 19/20:614–615.

Question 7: How do bats avoid predators?

Boinski, S., and R. M. Timm. 1985. "Predation by Squirrel Monkeys and Double-Toothed Kites on Tent-Making Bats." *American Journal of Primatology* 9:121–127.

Fenton, M. B. 1992. *Bats.* New York: Facts On File.

Foster, M. S. 1992. "Tent Roosts of Macconnelli's Bat (*Vampyressa macconnelli*)." *Biotropica* 24:447–454.

Kunz, T. H., and L. F. Lumsden. 2003. "Ecology of Cavity and Foliage Roosting Bats." In *Bat Ecology*, ed. T. H. Kunz and M. B. Fenton. Chicago and London: University of Chicago Press.

Tuttle, M. D. 2005. *America's Neighborhood Bats.* 2nd rev. ed. Austin: University of Texas Press.

Chapter Seven: Bats and People

Question 1: Why are people afraid of bats?

Benson, E. P. 1991. "Bats in South American Folklore and Ancient Art." *Bats* 9:7–10.

Cavendish, R., ed. 1970. *Man, Myth and Magic.* New York: Marshall Cavendish.

Kellert, S. R., and E. O. Wilson. 1993. *The Biophilia Hypothesis.* Washington, D.C.: Island Press.

Kern, S. J. 1988. "Bats in Chinese Art." *Bats* 9:13.

McCracken, G. 1992. "Bats and Human Hair." *Bats* 10:15–16.

———. 1992. "Bats in Magic, Potions, and Medicinal Preparations." *Bats* 10:14–16.

McCracken, G. 1993. "Bats and the Netherworld." *Bats* 11:16–17.

———. 1993. "Bats and Vampires." *Bats* 11:14–15.

———. 1993. "Folklore and the Origin of Bats." *Bats* 11:11–13.

Tuttle, M. D. 2005. *America's Neighborhood Bats.* 2nd rev. ed. Austin: University of Texas Press.

Question 2: What dangers do bats face from people?

Arnett, E. 2006. "Seeking Solutions for Wind Energy." *Bats* 24:1–6.

Baerwald, E. F., G. H. D'Amours, B. J. Klug, and R.M.R. Barclay. 2008. "Barotrauma Is a Significant Cause of Bat Fatalities at Wind Turbines." *Current Biology* 18:R695–R696.

Bat Conservation International. 2001. *Bats in Eastern Woodlands.* Austin, Tex.: Bat Conservation International.

French, B. A., L. Finn, and M. Kiser. "Bats in Buildings: An Information and Exclusion Guide." On the Web site Bat Conservation International. http://www.batcon.org/home/index.asp?idPage=51&idSubPage=48; accessed on March 25, 2008.

Tuttle, M. D. 2005. *America's Neighborhood Bats.* 2nd rev. ed. Austin: University of Texas Press.

Tuttle, M. D., and D.A.R. Taylor. 1998. *Bats and Mines.* Austin, Tex.: Bat Conservation International.

Question 3: Are bats useful to plants?

Fenton, M. B. 1992. *Bats.* New York: Facts On File.

Kalka, M. B., A. R. Smith, and E.K.V. Kalko. 2008. "Bats Limit Arthropods and Herbivory in a Tropical Forest." *Science* 320:71.

Kelm, D. H., K. R. Wiesner, and O. von Helversen. 2008. "Effects of Artificial Roosts for Frugivorous Bats on Seed Dispersal in a Neotropical Forest Pasture Mosaic." *Conservation Biology* 22:733–741.

Kunz, T. H., and M. B. Fenton. 2003. *Bat Ecology.* Chicago: University of Chicago Press.

Shilton, L. A., et al. 1999. "Old World Fruit Bats Can Be Long-Distance Seed Dispersers through Extended Retention of Viable Seeds in the Gut." *Proceedings of the Royal Society B: Biological Sciences* 266:219.

Thomas, D. W. 1991. "On Fruits, Seeds, and Bats." *Bats* 9:8–13.

Tuttle, M. D. 1986. "Gentle Fliers of the African Night." *National Geographic* 169:540–558.

———. 2005. *America's Neighborhood Bats.* 2nd rev. ed. Austin: University of Texas Press.

Williams-Guillen, K., I. Perfecto, and J. Vandermeer. 2008. "Bats Limit Insects in a Neotropical Agroforestry System." *Science* 320:70.

Question 4: Do people eat bats?

Borenstein, A. R., et al. 2007. "Cyad Exposure and Risk of Dementia, MCI, and PDC in the Chamorro Population of Guam." *Neurology* 68:1764–1771.

Cox, P. A., and O. W. Sacks. 2005. "Cyad Neurotoxins, Consumption of Flying Foxes, and ALS-PDC Disease in Guam." *Neurology* 58:956–959.

Friedland, R. P., and C. Armon. 2007. "Tales of Pacific Tangles." *Neurology* 68:1759–1761.

Galasko, D., et al. 2007. "Prevalence of Dementia in Chamorros on Guam." *Neurology* 68:1772–1781.

Hermosura, M. C., et al. 2005. "A TRPM7 Variant Shows Altered Sensitivity to Magnesium That May Contribute to the Pathogenesis of Two Guamanian Neurodegenerative Disorders." *Proceedings of the National Academy of Science* 102:11510–11515.

Morris, H. W. 2004. "Genome-Wide Analysis of the Parkinsonism-Dementia Complex of Guam." *Archives of Neurology* 61:1889–1897.

Plato, C. C., et al. 2002. "ALS and PDC of Guam: A Forty-Year Follow-Up." *Neurology* 58:765–773.

Sundar, P. D., et al. 2007. "Two Sites in the MAPT Region Confer Genetic Risk for Guam ALS/PDC and Dementia." *Human Molecular Genetics* 16:295–306.

Tuttle, M. D. 1983. "Can Rain Forests Survive Without Bats?" *Bats* 0:1–2.

———. 2005. *America's Neighborhood Bats.* 2nd rev ed. Austin: University of Texas Press.

Question 5: Do bats have any scientific or medical value?

Csontos, P., and P. Tatai. "Superbat—A Navigational Aid for the Blind." On the Web site National Office for Research and Technology. http://www.nkth.gov.hu/letolt/k+f/kf_angol/elements/PDF/IKTA5-050-NavigationAidForTheBlind-Superbat-Csontos.pdf; accessed March 2, 2008.

Hedenstrom, A., et al. 2007. "Bat Flight Generates Complex Aerodynamic Tracks." *Science* 316:894–897.

Locke, R. 2003. "The Vampire's Gift: Bat Saliva Yields a Promising Treatment for Stroke Victims." *Bats* 21:11–12.

Muijres, F. T., et al. 2008. "Leading-Edge Vortex Improves Lift in Slow-Flying Bats." *Science* 319:1250–1253.

Roach, J. 2002. "U.S. Navy Looks to Bats, Dolphins for Better Sonar." On the Web site National Geographic News. http://news.nationalgeographic.com/news/2002/12/1212_021212_batsonar.html; accessed March 2, 2008.

Stiles, E. 2008. "Research Focuses on Building 'Smart' Radar Systems." On the Web site University of Arizona News. www.uanews.org/node/18261; accessed March 2, 2008.

Question 6: Are any bats endangered?

McClintock, J. "Twenty Species We Might Lose: And Then There Were None." On the Web site Discover. http://discovermagazine.com/2000/oct/featspecies; accessed March 21, 2008.

Myers, N. 1979. *The Sinking Ark: A New Look at the Problem of Disappearing Species.* New York: Readers Digest Young Families.

Novacek, M. 2007. *Terra: Our 100-Million-Year-Old Ecosystem and the Threats That Now Put It at Risk.* New York: Farrar, Straus and Giroux.

Tuttle, M. D. "Gentle Fliers of the African Night." On the Web site Encarta. http://encarta.msn.com/sidebar_761593540/african_bats.html; accessed March 21, 2008.

Whitaker, J. O., Jr., et al. 2007. *Bats of Indiana.* Terra Haute: Indiana State University. Publication no. 1.

Wilson, E. O. "Vanishing before Our Eyes." On the Web site Time.com. http://www.time.com/time/reports/earthday2000/biodiversity01.html; accessed February 24, 2008.

Question 7: How can bats be safely evicted from a building?

French, B. A., L. Finn, and M. Kiser. "Bats in Buildings: An Information and Exclusion Guide." On the Web site Bat Conservation International. http://www.batcon.org/home/index.asp?idPage=51&idSubPage=48; accessed March 25, 2008.

McAney, C. 1992. *Bats and Bridges: A Report on the Importance of Bridges to Bats.* Galway, U.K.: Vincent Wildlife Trust Project.

Perlmeter, S. I. 1995. "Bats and Bridges: Patterns of Night Roost Use by Bats in the Willamette National Forest." *Bat Research News* 36:30–31.

Smiddy, P. 1997. "A Survey of Bats and Bridges." *Bat Research News* 38:37–38.

Tuttle, M. D. 2005. *America's Neighborhood Bats.* 2nd rev. ed. Austin: University of Texas Press.

Tuttle, M. D., and B. Keeley. 2001. *The Texas Bats and Bridges Project: Report to the Texas Department of Transportation.* Austin, Tex.: Bat Conservation International.

Question 8: How can I build a bat house?

Richardson, P. 2002. *Bats.* Washington, D.C.: Smithsonian Institution Press.

Tuttle, M. D. 2005. *America's Neighborhood Bats.* 2nd rev. ed. Austin: University of Texas Press.

Tuttle, M. D., M. Kiser, and S. Kiser. 2004. *The Bat House Builder's Handbook.* Rev. ed. Austin, Tex.: Bat Conservation International.

Question 9: How do scientists capture bats so they can study them?

Francis, C. M. 1989. "A Comparison of Mist Nets and Two Designs of Harp Traps for Capturing Bats." *Journal of Mammalogy* 70:865–870.

Gardner, J. E., et al. 1989. "A Portable Mist Netting System for Capturing Bats with Emphasis on *Myotis sodalis* (Indiana bat)." *Bat Research News* 30:1–8.

Kunz, T. H. 1988. *Ecological and Behavioral Methods for the Study of Bats.* Washington, D.C.: Smithsonian Institution Press.

Lollar, A., and B. Schmidt-French. 2002. *Captive Care and Medical Reference for the Rehabilitation of Insectivorous Bats.* 2nd ed. Mineral Wells, Tex.: Bat World Publication.

Palmeirim, J. M., and L. Rodrigues. 1993. "The 2-Minute Harp Trap for Bats." *Bat Research News* 34:60–64.

Tuttle, M. D. 1974. "An Improved Trap for Bats." *Journal of Mammalogy* 55:475–477.

Wilson, D. E., et al. 1996. *Measuring and Monitoring Biological Diversity: Standard Methods for Mammals.* Biological Diversity Handbook Series. Washington, D.C.: Smithsonian Institution Press.

Question 10: Can bats be domesticated?

Lollar, A., and B. Schmidt-French. 2002. *Captive Care and Medial Reference for the Rehabilitation of Insectivorous Bats.* 2nd ed. Mineral Wells, Tex.: Bat World Sanctuary.

Page, R. 2004. "Little Old Man: A Tale of Frogs, Fish and Amazing Memories." *Bats* 22:1–2.

Page, R., and M. Ryan. 2006. "Social Transmission of Novel Foraging Behavior in Bats: Frog Calls and Their Referents." *Current Biology* 16:1201–1205.

Question 11: How can I photograph bats?

Altenbach, J. S. 1988. "Techniques for Photographing Bats." In *Ecological and Behavioral Methods for the Study of Bats,* ed. T. H. Kunz, 125–140. Washington, D.C.: Smithsonian Institute Press.

Burian, P. K., and R. Caputo. 2003. *The National Geographic Photography Field Guide.* 2nd ed. New York: National Geographic Society.

Gerlach, J., and B. Gerlach. 2007. *Digital Nature Photography: The Art and the Science.* St. Louis, Mo.: Focal Press.

Hatasaka, M. 2004. *Digital Nature Photography: Use Any Digital Camera to Take Spectacular Nature Photographs.* Richardson, Tex.: Liberty Technologies Corp.

Tipling, D. 2007. *Digital Wildlife Photography.* Buffalo, N.Y.: Firefly Books.

Tuttle, M. D. "Photographing the Secret World of Bats." On the Web site Bat Conservation International. http://www.batcon.org/home/index.asp?idPage=88&idSubPage=33; accessed March 2, 2008.

Question 12: What is a bat detector?

Herman, J., et al. 2002. "Choosing a Bat Detector: Theoretical and Practical Aspects." In *Bat Echolocation Research: Tools, Techniques and Analysis,* ed. R. Mark Brigham et al., 28–37. Austin, Tex.: Bat Conservation International.

Sidebar: Bat Breath

"Bat Breath Reveals the Identify of a Vampire's Last Victim." On the Web site Science Daily. http://www.sciencedaily.com/releases/2007/08/070818110448.htm; accessed August 1, 2008.

Voight, C., et al. 2008. "Bat Breath Reveals Metabolic Substrate Use in Free-Ranging Vampires." *Journal of Comparative Physiology B: Biochemical, Systemic, and Environmental Physiology* 178:9–16.

Sidebar: Batman

"Batman." On the Web site The Great Batman Equipment Archive. http://members.fortunecity.com/kainwind/archive.html; accessed September 4, 2008.

Cotta, M. 1989. *Tales of the Dark Knight: Batman's First Fifty Years, 1939–1989.* New York: Ballantine Books.

Sidebar: Folklore

Benson, E. P. 1991. "Bats in South American Folklore and Ancient Art." *Bats* 9:7–10.

Kern, S. J. 1988. "Bats in Chinese Art." *Bats* 9:13.

McCracken, G. 1992. "Bats and Human Hair." *Bats* 10:15–16.

———. 1992. "Bats in Magic, Potions, and Medicinal Preparations." *Bats* 10:14–16.

———. 1993. "Bats and the Netherworld." *Bats* 11:16–17.

———. 1993. "Bats and Vampires." *Bats* 11:14–15.

———. 1993. "Folklore and the Origin of Bats." *Bats* 11:11–13.

Index

About the Authors

Barbara A. Schmidt-French is a biologist and the science officer at Bat Conservation International. She is also a wildlife rehabilitator and cares for dozens of injured and orphaned bats each year. She has authored both popular articles and scientific papers about bats and is the co-author of *Captive Care and Medical Reference for the Rehabilitation of Insectivorous Bats.*

Carol A. Butler, Ph.D., is the co-author of the Rutgers University Press question-and-answer series, including *Do Butterflies Bite?* (2008). She also co-authored *Salt Marshes: A Natural and Unnatural History* (2009), and *The Divorce Mediation Answer Book* (1999). She is a psychoanalyst and a mediator in private practice in New York City, an adjunct assistant professor at New York University, and a docent at the American Museum of Natural History.